# Open Innovation

## A Multifaceted Perspective

Part I

## Open Innovation: Bridging Theory and Practice

**ISSN 2424-8231**



---

*Published*

Vol. 1 Open Innovation: A Multifaceted Perspective (In 2 Parts)
*edited by Anne-Laure Mention and Marko Torkkeli*

*Forthcoming*

Open Innovation: Unveiling the Power of the Human Element
*edited by Anne-Laure Mention and Dimitrios G Salampasis*

Digital Innovation: Harnessing the Value of Open Data
*edited by Anne-Laure Mention and Vera J Lipton*

Open Innovation: Bridging Theory and Practice Volume 1

# Open Innovation

## A Multifaceted Perspective

Part I

Editors

Anne-Laure Mention
Luxembourg Institute of Science and Technology, Luxembourg

Marko Torkkeli
Lappeenranta University of Technology, Finland

World Scientific

NEW JERSEY • LONDON • SINGAPORE • BEIJING • SHANGHAI • HONG KONG • TAIPEI • CHENNAI • TOKYO

*Published by*

World Scientific Publishing Co. Pte. Ltd.
5 Toh Tuck Link, Singapore 596224
*USA office:* 27 Warren Street, Suite 401-402, Hackensack, NJ 07601
*UK office:* 57 Shelton Street, Covent Garden, London WC2H 9HE

**British Library Cataloguing-in-Publication Data**
A catalogue record for this book is available from the British Library.

Library of Congress Control Number: 2015953461

**Open Innovation: Bridging Theory and Practice — Vol. 1**
**OPEN INNOVATION**
**A Multifaceted Perspective**
**(In 2 Parts)**



ISBN 978-981-4719-17-9 (Set)
ISBN 978-981-4719-20-9 (Part 1)
ISBN 978-981-4719-21-6 (Part 2)

In-house Editors: Sutha Surenddar/Alisha Nguyen

Typeset by Stallion Press
Email: enquiries@stallionpress.com

Printed in Singapore by B & Jo Enterprise Pte Ltd

# FOREWORD

*Open Innovation: Bridging Theory and Practice* is an entirely appropriate and timely book that focuses on a globally pervasive current theme, and draws on inputs from manifold perspectives: academia, industry, practice and policy-making. Multidisciplinary is reflected by the plurality of the contributions, with diverse research methods, and a range of contemporary themes. The potential impact of this book, and the book series in general, lies in the explicit focus on open innovation as a recently emergent paradigm that has pushed us all beyond the traditional intrafirm innovation perspective.

Furthermore, the deliberate theme of the book series, "*Bridging Theory and Practice*" encourages authors and readers alike to expect, pursue, and peruse content that is coherent, topical, innovative, and useful. The bridging theme further encourages a focus of more traditional academic researchers to view the open innovation phenomenon through the lens of outcomes rather than inputs and activities, and take into account the real-world hurdles that modern organisations face in the dynamic, gruelling competitive struggle for survival and growth: How to draw on their ecosystem to meet the rapidly changing unmet needs of the customer, how to keep up with the unprecedented pace of change in technological trajectories, how to adapt to political upheavals, how to adjust to shifts in legal and regulatory boundaries frameworks, how to attend to dynamic cultural differences — all these within a global socioeconomic milieu characterised by volatility, instability and shifting of global forces, but at the same time perceiving and exploiting emerging opportunities.

The content of the first two books, along with the themes discussed throughout the book series, suggest that open innovation is a pilgrimage — is indeed a journey with many challenges that require creativity, imagination, insightful thinking, collaboration and vision for those enterprises that would pursue innovation.

So, this book series will become a must-read resource for researchers, students, academic scholars, innovation practitioners, industry leaders, journalists, and policy-makers, in fact every member of the innovation ecosystem who wishes to keep pace with the latest thinking about open innovation as a driver of growth and prosperity.

The series will help readers learn from the past, exploit the present and seize the future, deploying open innovation by entwining theory and practice. So, let the journey begin ...

Ian C. MacMillan
The Wharton School

# ABOUT THE EDITORS

**Dr. Anne-Laure Mention** is holding an expert position at LIST. Until end of 2014, she was leading a research unit focusing on innovation economics and management within the Public Research Centre Henri Tudor, Luxembourg. She is actively involved in research projects, mainly focusing on innovation and performance measurement and management in the financial and business-to-business services industries. She is also currently appointed as Expert to the High Committee for the Financial Industry in Luxembourg, in the context of the FinTech initiative. She also has a Visiting Professorship position in University of Liège, Belgium and is the Deputy Director of the Centre d'Evaluation de la Performance des Entreprises.

Her research interests mainly concentrate on open and collaborative innovation, intellectual capital measurement and management, innovation and technology management, with a focus on financial and business-to-business services industries. She has published in leading academic journals and has been a Visiting Researcher at McGill University, Canada, at Ferrara University, Italy and at Singapore Management University. She received two prestigious IBM Faculty Awards for her works in the field of Open Innovation in services, is a founding member of WICI, and the Deputy Head of the ISPIM Advisory Board. She is one of the founding editors of the open access, multidisciplinary *Journal of Innovation Management*. She is regularly involved as a Jury Member in international competitions revolving around financial innovation and has gained practical experience in the consulting and venture capital industries.

She is the co-editor of a book published by Cambridge Scholars Publishing, entitled "Innovation in Financial Services: A Dual Ambiguity".

**Dr. Marko Torkkeli** is a Professor of Technology and Business Innovations at the Lappeenranta University of Technology, Finland.

His research interests focus on technology and innovation management, strategic entrepreneurship, growth venturing, and decision support systems. He has published over 200 articles in academic journals. Dr. Torkkeli has also been engaged in numerous international research projects in the USA, Australia, Brazil, Canada, China, India, Japan, Russia and across Europe. He is a Visiting Researcher at INESC Porto, Portugal, a Docent of Technology-based Business at University of Jyväskylä, Finland, a Docent of Technology and Innovation Management at Helsinki University of Technology, Finland and holds an Affiliated Faculty position at Singapore Management University. He has more than 15 years of experience of consulting activities in innovation management and strategy. He serves as the Director of Publications of the International Society for Professional Innovation Management (ISPIM) and is one of the founding editors of the open access, multidisciplinary *Journal of Innovation Management*.

He is the co-editor of a book published by Cambridge Scholars Publishing, entitled "Innovation in Financial Services: A Dual Ambiguity".

# LIST OF CONTRIBUTORS

**Tor Helge Aas** is an Associate Professor at Department of Management, School of Business and Law, University of Agder, Norway. Dr. Aas obtained his PhD in strategy and management from the Norwegian School of Economics (NHH) in 2010. He is conducting research in innovation management, management control and strategic management, particularly in relation to the service sector. His research concentrates on topics such as innovation strategy, management of innovation processes, management control of innovation activities and innovation collaboration (including open innovation, high involvement in innovation and innovation systems).

**Frederic Adam** is a Professor and the Director Master of Business Studies in Innovation in European Business and the Head of the Graduate School of the College of Business and Law, University College Cork, Ireland. Other positions he holds include: Principal Investigator in the Financial Services Innovation Centre (FSIC), Editor-in-Chief of the *Journal of Decision Systems* and the Chair of the Working Group 8.3 on DSS of the International Federation for Information processing (IFIP). He holds Doctorates from the National University of Ireland and Université Paris VI (France). His research interests are in decision-making and decision support and in ERP. He has over 25 journal papers published in international journals including *Information and Management, The Journal of*

*Strategic Information Systems*, *Decision Support Systems*, and the *Journal of Information Technology*. He held visiting Research fellow positions at the London School of Economics (2005–2007) and at Lund University, Sweden (2009–2011) and has successfully supervised 8 PhD students to completion.

**Pieter Ballon** is a professor at iMinds-SMIT (VUB). Pieter specialises in business modelling, open innovation and the mobile telecommunications industry. Formerly, he was senior consultant and team leader at TNO. In 2006–2007, he was the coordinator of the cross issue on business models of the Wireless World Initiative (WWI), that united 5 Integrated Projects in the EU 6th Framework Programme. Currently, he is also the international Secretary of the European Network of Living Labs. Pieter Ballon holds a PhD in Communication Sciences and a MA in Modern History.

**Marcin Baron** (Ph.D. Econ.) is an Assistant Professor at University of Economics in Katowice, Poland (Department of Strategic and Regional Studies) and partner at InnoCo Ltd. (business consultancy, Poland). Graduate of University of Economics in Katowice and Foreign Commonwealth Office scholar in Oxford. Combines academic theory with over 10-year expertise in innovation and regional policy as well as innovation management. Currently focused on industrial policy, smart specialisations and open innovation. Skilled in foresight techniques and strategic management tools. Experienced project manager and researcher, active in international teams and regional platforms. Joins social activities related to leadership, creativity and entrepreneurship. Interested in photography and alpine skiing.

**Karl Joachim Breunig** is an Associate Professor at the Business School of Oslo and Akershus University College of Applied Science, Norway. Currently, Karl Joachim Breunig is receiving his PhD in Strategic Management from BI Norwegian Business School and holds an MSc from London School of Economics. His research concentrates on knowledge work and innovation, in particular in relation to services. He pioneered the knowledge management profession in Norway as

Knowledge Manager in an international telecommunications consultancy, and was Co-founder of the Knowledge Management Forum in Norway, serving as its first President. He has been engaged in research projects on knowledge management, innovation, and internationally distributed knowledge work in a number of both public-sector organisations and private organisations in several different industries.

**Fergal Carton** is a Senior Researcher in the FSIC at University College Cork, Ireland. Fergal's research domain is the integration of information technologies in financial services, with a specific focus on the interface between transactional data and decision support. With 15 years' experience as a management consultant, starting with the Boston Consulting Group in London, he has worked on over 30 projects involving the definition of IT strategy, enterprise systems integration and business model development. With a primary degree in Computer Science (University College Dublin), and an MSc in Management from the European School of Management Studies (ESCP-EAP) in Paris, Fergal's doctoral research focused on the value of enterprise-wide systems in supporting management decision-making. Fergal spearheads FSIC research activities relating to payments and specifically the integration of mobile payment into mainstream financial services.

**Leonid Chechurin** received his Doctor of Technical Science Degree in 2010 with a dissertation on Mathematical Modeling and Analysis of Dynamic Systems. His work has been published in journals and conferences in the fields of control system theory and automation, mathematical modeling, creativity and innovations. He has more than 40 publications in these fields. He has been involved in the supervision of about 50 M.Sc. theses and dissertations. At the time of writing the chapter, Chechurin was employed by Samsung Electronics as Principal Engineer. Currently, he works as a Professor at Lappeenranta University of Technology. Leonid was employed by leading innovating technology companies like Samsung SDI, Samsung Electronics and LG Electronics as a consultant/principal researcher for the engineering design group (5 years in total). As a part of Gen 3 Partners company, he had various teaching and mentorship assignments (in total more than 50 seminars and consulting sessions and several research projects on inventive engineering design). Prof. Chechurin was the Visiting Professor at KumOh National University of Technology (South Korea, 1998–1999) and the Fellowship Professor at Politecnico di Milano (Italy, 2011). Currently, he is the Fellow Researcher for Yakutsk State University named after Ammosov. In 2004–2012, he served as the Head of the Department (Theory of Innovatics) at St. Petersburg State Polytechnical University, where he currently holds the position of Research Fellow. Prof. Chechurin is member of the Russian and international academic communities and editorial boards, Vice-President of MATRIZ, International Association of TRIZ (Theory of Inventive Problem Solving).

**Lieven De Marez** is Head of the iMinds-MICT Research Group at Ghent University and Manager of iMinds iLab.o's media living lab activities, where he continuously seeks to explore new methodologies. Main goals are understanding emerging media use patterns and the impact of new media & ICT and making media innovation more user-centric. At the Department of Communication Studies, he founded and coordinates the Master 'New Media & Society'. His main expertise is on the development of 'segmentation forecast' tools for prior-to-launch adoption potential forecasts for new media and ICT-innovations.

**Denis Dennehy** obtained his PhD in 2014 with the Business Information Systems Department, University College Cork, Ireland. His research adopted a design science approach which led to the design of a visualisation tool which enables multiple partners to manage key partnership issues when engaging in an m-payment value network. He won the Best Academic Paper Award at the 2013 Innovation for Financial Services Conference in Singapore and in 2012, he received the Marion McAneney Graduate Research Award; a national award that acknowledges excellence in reviewing academic literature on any aspect of innovation and identifying gaps in our understanding. Prior to this, he completed a research masters which explored the gap between policy and practice in international development which was motivated by his work in a developing country. He has worked on a number of industry focused research projects and his published topics include: Design thinking; design science; mobile payments; knowledge management and international development.

**Antti Hautamäki** now a Professor *Emeritus*, was in the years 2009–2013 a Research Professor and the Director of the Agora Center at the University of Jyväskylä. His research field was innovation processes and service innovation. He has conducted several research projects related to innovation ecosystems, local innovation hubs and innovation policy. In 1996–2008, he was a Director of research and innovation program at the Finnish Innovation Fund Sitra. He is also an adjunct professor of theoretical philosophy at the University of Helsinki. In 2006–2007, he was a visiting scholar at UC Berkeley, School of Information. Dr. Hautamäki has published and edited about 30 books and published 200 articles about philosophy, cognitive science, innovation and information society. His latest books include *Sustainable Innovation, A New Age of Innovation and Finland's Innovation Policy*, Sitra reports 87, 2010, *Towards innovation center* [in Finnish] with Kaisa Oksanen, 2012, and *Contradictory Science Policy* [in Finnish] with Pirjo Ståhle, 2012. Currently, Dr. Hautamäki is working in his firm Consulting Sustainable Innovation.

**Katja Maria Hydle** is a Senior Research Scientist at IRIS Social Science. She earned her political science education in Switzerland and Belgium and is completing her PhD in Strategy from BI Norwegian Business School. She has extensive research experience with service providers both nationally and internationally. Hydle has a broad interest in service work, strategy and innovation and has worked in large research projects on transnational service provision, collaboration practices and service innovation. She has published in these topics in international peer-review journals, conferences and books.

**Jari Kaivo-oja** is a Research Director of Finland Futures Research Centre (Turku School of Economics). He works currently in the CHEC project (with CASS, China) and in some smaller R&D projects. He has specialised in foresight methodology, innovation research and sustainability studies. Jari Kaivo-oja has a doctoral degree in administrative sciences (regional and environmental sciences) from the University of Tampere. He has also Master Degree in International Economics. He is Adjunct Professor at the Faculty of Science at University of Helsinki (Faculty of Science, Geosciences and Geography, Planning Sciences) and at the Faculty of Social Sciences at the University of Lapland (Innovation and Foresight Research). Jari Kaivo-oja is expert working in the fields of foresight, planning sciences and innovation research. His special fields are foresight methodologies in sustainability analyses, international energy economics and general societal development. He has special expertise in energy economics, innovation research, climate change analyses and security politics. He has broad research experience in strategic and participatory foresight processes in Europa and in international contexts. He has worked in the European FP6 and FP7 research programs. He has worked as a Finnish national co-ordinator in the COST A22 (Foresight methodologies) and as an expert member of the international think tank of European Sustainability Strategy Theme Network. He has worked for the European Foundations (Dublin) and for the European Parliament (Brussels), as well as for the Nordic Innovation Centre (NIC, Oslo). In Finland, he has been a researcher of the Academy of Finland, the VTT (Technical Research Centre of Finland), the Finnish Technology and Innovation Fund TEKES and various line ministries.

**Miika Kajanus** works currently as a Coordinator and Project Leader at the Entrepreneurship & Innovation cross-cutting profile area in Savonia University of Applied Sciences. His main tasks are to organise innovation processes with enterprises and other organisations as a part of regional development strategies. The tasks contain connecting scientific expertise related to the societal challenge to a multi-disciplinary perspective, open to the further cross-fertilisation of ideas between disciplines and sectors. The work is collaboration mainly with SMEs, inventors, end-users, students, experts, investors and regional development authorities aiming in a straightforward way to implement innovations including all the phases starting from idea generation, conceptualising, business models development and commercialisation. In that position, Dr. Kajanus has been working since 2009 and before that as a Research & Innovation Manager in Ylä-Savo Region since 2004. He has published more than 100 successfully implemented innovations into markets with SMEs during 2008–2012 and about 20 scientific articles.

**Bram Lievens** is a Senior Researcher at iMinds-SMIT (VUB). Bram is responsible for the Living Lab research track within the user research unit. Bram's main expertise is within the domain of user-centered design and Living Labs, with a focus on mobile applications in a smart city environment. Bram is active in and supervising various research projects in this area.

**Vera Lipton** is a PhD candidate at Thomas More Law School at the Australian Catholic University. As an advocate of open innovation and an intellectual property (IP) counsel, she has dedicated much of her recent time and work to the law governing technology, science and the Internet, with primary focus on the interface between shared and exclusive knowledge. She is experienced in contributing empirical input to high-level policy discussions on open science, IP and data management, research impact assessment, and university–industry collaborations. Previously, Vera founded the National Patent Analytics Hub at IP Australia, worked as an executive Australian Government officer, and, before that, as a consultant for an international technology transfer company. Currently, she lives in Cyprus and works internationally.

**Anne-Laure Mention** is an Innovation Economics and Management Expert at the newly created Luxembourg Institute of Science and Technology, the Deputy Director of the Centre D'Evaluation de la Performance des Entreprises at University of Liege, Belgium, and

Visiting Professor at the University of Liege, Belgium. Prior to joining LIST, she was leading a research unit focusing on innovation economics and management within the Public Research Centre Henri Tudor, Luxembourg. Her research interests mainly concentrate on open and collaborative innovation, intellectual capital measurement and management, innovation and technology management, with a focus on financial and business-to-business services industries. She has published in leading academic journals and has been a Visiting Researcher at McGill University, Canada, at Ferrara University, Italy and at Singapore Management University. She received two prestigious IBM Faculty Awards for her works in the field of Open Innovation in services, is a founding member of WICI, and the Deputy Head of the ISPIM Advisory Board. She is one of the founding editors of the open access, multidisciplinary *Journal of Innovation Management*. She is regularly involved as a Jury Member in international competitions revolving around financial innovation and has gained practical experience in the consulting and venture capital industries.

**Kaisa Oksanen** (PhD, Public and Social Administration) is a Senior Scientist in VTT Technical Research Centre of Finland. Previously, she worked as a Research Coordinator and Innovation Researcher at the Agora Center, University of Jyväskylä, in Aalto Universty and in the Finland Futures Research Centre. Her background is in social and political sciences, systemic innovation and democracy development. She took her PhD in the City University of Hong Kong (2005–2009) and in recent years she has been working with projects concerning e.g. regional development and innovation ecosystems, university-based innovation, structural change, service innovation, competitiveness and well-being.

**Jarkko Pellikka** holds a Ph.D (Economics and Business Administration) and currently works as Head of Operations in Nokia Labs Unit as a part of Nokia Technologies. He has worked several years in the leading global multinational companies and has been responsible for managing and developing numerous global business development initiatives in his field of expertise. His research focuses on commercialisation process of innovation, operations management and business development in technology industries. His research on commercialisation process of innovation, innovation management and business leadership in high technology industry has been published in several international journals and books.

**Teemu Santonen** received his PhD (Economics) degree in Information Systems Science from Aalto University in Finland in 2005. Currently, he is acting as a Principal Lecturer at the Laurea University of Applied Sciences. Within Laurea, he has lead various projects gaining 1.7 MEUR cumulative funding and award from the Finnish inventor support association for the best school-related innovation. Santonen has published over 40 articles in international peer-refereed journals and conferences and his research interest includes innovation management, foresight, business models, learning methods, cross/inter cultural collaboration, security management and social network analysis. Santonen is also a scientific panel member of ISPIM. Besides academic experiences, Santonen has more than a decade of practical consulting and development experience in leading Finnish financial, media- and ICT-sector organisations. In Laurea, Santonen has filed several invention disclosures, which have resulted in a start-up company and one patent.

**Bror Salmelin** is Policy Advisor to the Directorate General for Communications, Network, Content, and Technology (DG CONNECT) where he is responsible for Open Innovation and modern Innovation Systems, since 2011. In parallel, he runs a senior industrial group "Open Innovation Strategy and Policy Group" with leading industries. In 2007, he was Policy Advisor for the Director in ICT addressing Societal Challenges and he began his career at the European Commission in 1998 as Head of Unit. Bror Salmelin is Member of New Club of Paris and Member of the Advisory Board for Innovation Value Institute, Ireland. Bror Salmelin graduated from Helsinki University of Technology in 1978 and he was Assistant at Helsinki University of Technology from 1979–1984. Consequently, he worked for TEKES (a Finnish agency co-ordinating industrial RTD) throughout various management positions (e.g. in Manufacturing, Industrial Automation and Electronics) and in 1994, he was nominated Deputy of the Information Technology Section.

**Dimitri Schuurman** is a Senior Researcher at iMinds-MICT (Ghent University) and is working at iLab.o where he keeps an eye on the methodology of Media and ICT Living Labs. He is doing research about the role Living Labs play in Open Innovation, Smart Cities and User-Centric Innovation. More specifically, he focusses on user profiles, such as Lead Users and detectors, in innovation processes. Using his working-experience as a solid ground, Dimitri is currently finalising his PhD with the working title: 'Bridging the gap between open and user innovation: Living Labs as a means to structure user involvement and manage innovation'.

**Marko Seppänen,** PhD, Associate Professor in the field of technology management and innovation in the Department of Industrial Management, is Director of the Center of Innovation and Technology Research (CITER). CITER focuses on three broad areas: Modelling technological evolution, the co-evolution of technology, enterprises and industry structures, and technology strategy and innovation management. Dr. Seppänen has published more than 100 academic peer-reviewed papers, e.g. in journals such as *Journal of Product Innovation Management, Technological Forecasting and Social Change, Journal of High Technology Management Research, and International Journal of Physical Distribution & Logistics Management.* His current research interests are with technology commercialisation and business models.

**Marko Torkkeli** is a Professor of Technology and Business Innovations at the Lappeenranta University of Technology in Kouvola, Finland. His research interests focus on technology and innovation management, strategic entrepreneurship, growth venturing, and decision support systems. He has published articles in journals such as *International Journal of Production Economics,*

*International Journal of Foresight and Innovation Policy, International Journal of Business Excellence, International Journal of Technology Management and International Journal of Technology Intelligence and Planning*. He is a member of the editorial boards of the *International Journal of Services Sciences* and the *International Journal of Innovation Management*. He is a Visiting Researcher at INESC Porto, Portugal, a Docent of Technology-based Business at University of Jyväskylä, Finland, a Docent of Technology and Innovation Management at Helsinki University of Technology, Finland and Affiliated Faculty at Singapore Management University. He serves as the Director of Publications of the *International Society for Professional Innovation Management* (ISPIM) and is one of the founding editors of the *Journal of Innovation Management*.

**Carina Veeckman** is working in the User Research Department of iMinds-SMIT since 2011. She is active in different user-oriented projects; and her main expertise is within the domain of Living Labs. Until March 2013, Carina was responsible for the living lab methodology within the Flemish Living Lab Platform (FLELLAP), which included numerous projects within the smart grids, smart media, and smart cities domains with a test panel of 2,000 users. Her current research and interests are related to open data and the co-creation of mobile applications within a smart city context, and the willingness to share location information when using these applications.

# CONTENTS

# Chapter 1

# UNFOLDING FIFTY SHADES OF OPEN INNOVATION: STIMULATING INSIGHTS & FORESIGHTS

Anne-Laure Mention and Marko Torkkeli

Ὁ βίος βραχὺς, ἡ δὲ τέχνη μακρὴ, ὁ δὲ καιρὸς ὀξὺς, ἡ δὲ πεῖρα σφαλερὴ, ἡ δὲ κρίσις χαλεπή

(Hippocrates)

*Life is short, art is long, times are fleeting, experience is perilous and decision-making is difficult.*

Since the concept of Open Innovation was coined by Henry Chesbrough in 2003[1], it has increasingly gained popularity in academic research and attracted interest of practitioners and policymakers alike. A recent search (December 2014) performed in Scopus revealed a mere 5 hits in 2003, a more significant 161 hits in 2009 and over 300 hits in 2013, considering exclusively journals, conference papers, reviews and editorials listed in this internationally recognised database of peer-reviewed literature. From the business perspective, according to Google search, looking for "Open Innovation" in April 2003 returned about 200 hits, while performing the same search 10 years later provided over 450 million links.[2] It is obvious that Open Innovation became a paradigm which has attracted attention not only among academics, but also business entities and beyond. More and more people handing out their business cards show titles related to open innovation, like chief open innovation officer, open innovation specialist, head of open innovation to name a few. Job offerings website Monster.com search with the skill "Open Innovation" produced from tens to hundreds of job ads depending on your country of interest (conducted in March 2015). The debate about being open or not has reached new arenas and disciplines. For example, open data is booming where data owners (often public bodies) are wondering how much and under which conditions they can open-up databases and files. At the same time, the business world is eagerly looking for new opportunities to grow via open data which usually creates more jobs, desperately hunted by governments.

There has been evident development at policy level as well. The circle of Open Innovation will be closed when we add the policy perspective next to academia and practice. There is growing evidence that countries and policymakers have realised the untapped potential of Open Innovation. US President Obama supported and promoted the Open Government Initiative in 2009. The underlying idea of this act is to promote efficiency and effectiveness in government, while ensuring transparency, public participation and collaboration. Democratising publicly funded research and securing fair access to science has also been on the political agenda of the leading worldwide economy. The European Union has initiated similar actions through the launch of its initiative Open Innovation 2.0 based on a "Quadruple Helix Model where government, industry, academia and civil participants work together to co-create the future and drive structural changes far beyond the scope of what any one organisation or person could do alone" (European Commission Website). It has been considered as an important element in the European Innovation System. Several countries have their own Open Innovation initiatives going on. For example, the uncontested leading innovative economy in Europe, Finland, made their initial country-level research in 2008 about the implementation of Open Innovation in the country, focusing on how companies, universities and public sector interact and co-operate.[3] Most countries are following up, with similar initiatives.

Last but not least, another element of policymaking related to Open Innovation lies in the way we educate the next generation. Several universities have courses including Open Innovation elements but at this stage, cursus focused exclusively on Open Innovation have not been specifically developed. This has been the rationale for launching a large, European-wide project, funded under the Erasmus Academic Networks, which aims at elaborating a set of curricula for university students, ranging from bachelors to PhD students, and executive MBAs. Gathering 52 partners across Europe, this project (www.oi-net.eu) intends to pave the way for the development of structured and standardised curricula across Europe.

Despite the growing research efforts focusing on Open Innovation, there is still a vibrant debate on what Open Innovation actually entails and to what extent the concept is valuable to depict economic and managerial phenomena. This decade of Open Innovation research has undoubtedly unveiled significant findings on the concept, its perimeter, its adoption and has also paved the way for a detailed and rich research agenda. The existence of this debate and research agenda also pinpoints to the fact that critical and foundational questions remain unanswered and deserve further attention.

This book series "Open innovation: Bridging theory and practice" precisely aspires to respond to this research agenda, by answering to some foundational and fundamental questions that currently remain unaddressed. The series aims to contribute to knowledge creation and more importantly, to knowledge accumulation, through the combination of multiple streams, perspectives, disciplinary approaches and diverse backgrounds. In doing so, it departs from the current body of literature adopting a purely academic perspective on Open Innovation, and thus restates the importance of anchoring Open Innovation research into the reality, practices, challenges facing firms and policymakers. Its added value lies in the fact that it gives room not only to academics but also to innovation professionals to describe their Open Innovation journey, reflecting on their challenges and experiences in such a way that they provide meaningful insights to other practitioners, thought leaders and policymakers.

This book series gathers selected contributions from academics, thought leaders and policymakers active in the field of Open Innovation, across industries, sectors and countries. Its main originality stems from the fact that it astutely bundles theoretical insights with managerial practice, so as to provide topical insights to a wide readership, encompassing academics, researchers, innovation leaders, and policymakers with a shared interest in innovation. The attention granted simultaneously to managerial practice and policy-making is, in the view of the editors, an essential feature of this series, as it enables grasping industrial and sectoral peculiarities, diverse institutional contexts, and practical experiences in view of offering fresh, topical yet sustainable, multidisciplinary and multipolar perspectives on Open Innovation. This approach also aims at making sense out of a concept that, they believe, represents much more than a management fad by enrolling the wider community involved in the implementation of Open Innovation and benefitting from their insights. In other words, this series aims at having an impact where it matters the most: In firms and organisations, where the economic reality is happening.

This series covers multiple perspectives, such as measuring and assessing the impact of Open Innovation, dealing with organisational matters and culture, designing strategies, policies, incentives and measures to support and implement Open Innovation, and discussing the advantages and limitations of adopting Open Innovation strategies. Across the different books, it will deal with Open Innovation at individual and team levels, exploring the human side of Open innovation; at firm level, addressing the endless managerial challenges faced by organisations; at ecosystems level and finally at macroeconomic level, focusing on policy elaboration and impact analysis. A consistent feature of the book series resides in its willingness to provide topical insights on real cases of Open Innovation adoption in firms, while simultaneously contributing to, and hosting the ongoing debate on this paradigm shift.

The first two books of this series will exemplify our conviction that Open Innovation needs to be debated in different arena, from multidisciplinary perspectives and most importantly, with practical insights from thought leaders and policymakers. The contributions revolve around three main themes. First, we will address the links between Open Innovation and various disciplines (such as strategy, economics), methods (such as design thinking and futures), concepts (sustainable and symbiotic innovation) and policy instruments (living labs). This first part will also challenge the necessity of introducing the concept of "Open Innovation", as perceived in a provocative contribution by a thought leader in a large multinational company. Second, we will review selectively the literature, focusing essentially on open service innovation and on innovation in financial services industries. We will then explore different forms and types of practices reflecting the adoption and implementation of Open Innovation. Namely, we will explore co-creation processes, customer involvement, crowdsourcing, communities of practices, using different research methods (ethnographic, case study research and interviews) and combining views from academics and practitioners. Third, we will move towards the management of Open Innovation, paying specific attention to the individual, intra- and inter-organisational levels, as well networks as units of analysis. Managing culture and diversity will also be debated in this section, and will be illustrated using a European-wide case.

## Part 1: Open Innovation: Adopting a Multipolar Approach

Proudly stealing part of the famous quote from Hippocrates, we adopted "Art is long, and life is short" as our motto. This has been our motivation for launching this book series and for selecting a practitioner-oriented contribution, which explicitly and directly challenges the necessity of introducing the concept of Open Innovation, as well as the challenges related to its implementation in companies. This choice resonates with our wish to trigger the disturbing questions and to raise the practical concerns that are left unsolved in the heads of managers and policymakers alike. Our aim is to host this debate, by collectively involving and engaging a wide variety of actors. Specific attention has been paid to striking the right balance between academic contribution, with theoretical relevance, and the pragmatic insights that firms and policymakers still desperately need. A significant part of the chapters also astutely combines theoretical and practical views, and has been the outcome of cooperation between academics and practitioners.

In Chapter 2, "Open questions about Open Innovations", Leonid Chechurin, an academic turned practitioner who just went back to academia, raises the question of the relevance and usefulness of the concept of Open Innovation. Based on his experience as a practitioner for a leading industrial, with a strong R&D orientation, multinational company, Chechurin challenges the realistic nature of openness in the innovation process. Referring to some practical experience, the author argues that Open Innovation still has to evidence its suitability in practice, its novelty and its fruitfullness. He also emphasises some of the key individual and organisational challenges related to the adoption of Open Innovation strategies, such as corporate culture and intellectual property rights management.

Chapter 3 by Vera Lipton, "Open Data for Open Science: Aspirations, Realities, Challenges and Opportunities", opens the debate to all scientific fields. By advocating for open data and free flows in the scientific community, she embraces some of the key questions of this century: How to create new knowledge while not reinventing the wheel? How to benefit from existing knowledge to further expand the boundaries of our understanding of the world, to create radical novelties? How to build repositories of knowledge, and to make them publicly available? Vera Lipton recollects the history of open scientific data, since the emergence of the phenomenon. The Scholar builds the case for openness in science, arguing on the need to share scientific knowledge as widely and as freely as possible. As she puts it, "if science is to deliver its full value to society, it must be easily and freely accessible". Furthermore, she claims that open science is a key to increase the dissemination of research findings, to augment its impact on society, to reduce the duplication of research effort and finally, to enhance the quality of the scientific outcomes and methods thanks to quicker and faster feedback. She further provides an in-depth review of current initiatives in terms of open scientific data as well as the positioning of selected countries with respect to the open access policy. Her analysis follows with the evolution of publishing practices, and then conducts an in-depth exploration of the case of the CERN regarding its open data practices. She concludes with managerial and policy recommendations, which would ultimately increase the volume of re-usable scientific data publicly accessible.

Our journey continues with an academic contribution which concentrates on the linkages between strategy and innovation. But what is strategy without innovation? "Innovate or die" has been a prevailing mantra in businesses, but it seems more topical and relevant than ever in these turbulent times. Karl Joachim Breunig, Tor Helge Aas and Katja Maria Hydle examine the complex relationship between strategy and innovation in the context of open service innovations. Based on a multiple case analysis, embracing five large scale-intensive service firms belonging to several service industries, their results highlight the absence of explicit innovation strategies in all firms and suggest that the logic of innovation and strategy is hard to integrate in practice for these firms. Their empirical exploration contradicts the extant yet ambiguous literature on the links between strategy and innovation, and provides stimulating insights for managerial practice.

Sustainability now lies at the cornerstone of most strategies and policies, and analysing innovation through this prism brings insightful findings. Antti Hautamäki and Kaisa Oksanen introduce the concept of sustainable innovation, defined simultaneously by an impact orientation, a systemic and inclusive approach and a need to tap into undiscovered innovation potential and new sources of knowledge. They contend that the challenges facing the world nowadays call for a shift in the way innovation is defined, implemented and managed in firms and at policy level. Through a literature review, as well as an exploration of several examples, they build the case for sustainable innovation, which aims at balancing the long term influence of the process and the output with the needs of people and societies, as well as economic and environmental concerns.

After exploring what and how sustainability can make innovation thrive, we move forward to another strategic approach of Open Innovation, adopting a foresight or more precisely, a futures-oriented perspective to shed light on its relevance, and complementarity with the closed innovation model. In their academic chapter, Jari Kaivo-oja and Teemu Santonen explore the Open Innovation paradigm through the lenses of futures-oriented approaches of innovation management. Applying this perspective confirms that Open Innovation and Closed Innovation models are complementary and not substitutes for large firms, which apply both in a balanced and strategic way to augment their research and innovation capabilities. They introduce the concept of "symbiotic innovation management model" which falls somewhere between fully open and fully closed innovation models, thus striking the right balance, according to the needs, culture, and development goals of the firms. The scholars further unveil that global networking is a key driver for Open Innovation activities, thus suggesting that appropriate mechanisms, incentives and routines should be implemented so as to thrive the Open Innovation adoption. They also stress the role of technological change, and their overall turbulence levels, to drive the adoption of Open Innovation practices. Finally, they conclude that radical innovation represent the key success factor of growth in companies and suggest to open the debate on the integration of theories of radical innovation to the newly developed Open Innovation paradigm.

We temporarily close the conceptual debate, and will reopen it in every book of this series, to move forward to more tactical levels, thus introducing tools, methods, techniques which may cleverly be integrated into an Open Innovation approach. Design Thinking and Business Process Management will be the focus of those tactical contributions, bringing about more clarity for innovation leaders, as well as some concrete and pragmatic toolkit, and experience feedback on those.

Denis Dennehy, Frederic Adam and Fergal Carton debate about the relevance of design thinking as a new approach to solve complex problems and challenges in businesses, when conducting activities aimed at the development of novelties or when elaborating new strategies. Their contribution depicts how design thinking has been applied for a wide range of industries in two small and open economies, and elaborates on the benefits perceived by practitioners, when confronted with design thinking approach and visualisation tools such as the Partnership Management Canvas®.

Jarkko Pellikka, Miika Kajanus and Marko Seppänen examine the inter-organisational collaboration as an essential practice of Open Innovation adoption in the global process industry. The scholars define, test and implement a standardised process, based on Business Process Management and Lean Six Sigma approaches, as well as evaluation methods and tools that engage several collaborators to improve key business process performance of their firms. Their findings demonstrate that adopting a standardised process can lead to several concrete business results for the management, as well as can foster the adoption of Open Innovation. Finally, their multiple case analysis illustrates that engaging internal stakeholders acts as a catalyst for the utilisation of the firm's tacit knowledge to further improve business performance. Interestingly, their multiple case study tackles several levels of analysis and provides stimulating insights on how fostering inter-organisational knowledge sharing among the collaborators facilitates the creation of a variety of novelties at the firm, regional and cluster levels.

Departing from the firm level, or inter-firm level of analysis, we then turn to public–private partnerships and more specifically living labs. Dimitri Schuurman, Bram Lievens, Carina Veeckman, Lieven De Marez and Pieter Ballon exemplify, with the case of a Living Lab, the distributed innovation process via a collaboration between multiple partners through a public–private partnership where the end users play a dominant and primary role. Based on their exploration of an ICT living lab, through the lenses of Open Innovation and innovation networks and systems, the authors unveil the knowledge transfers, constellation and orchestration taking place in the living lab under scrutiny. They conclude with some recommendations in terms of governance for such forms of open and collaborative space for innovation.

Moving along the scale of unit of analysis, we now reach the regional level with a more economics-oriented approach to Open Innovation. In his contribution, Marcin Baron explores the links between Open Innovation and territory, through an in-depth analysis of territorially anchored initiatives fostering Open Innovation. The author first develops the key features of Open Innovation versus those of Territorial Innovation, explored through the lenses of their orientation, objective and main underlying notions. He further reviews to what extent Open Innovation is present in territorial policies at European level, mainly in the context of the European Cohesion Policy 2014–2020 investment priorities, and then concentrates on how these thematic priorities have been addressed in selected recommendations for all European countries in the Commission's policy papers. Ultimately, the author raises the question of the possible complementarity between the underlying concepts of Open Innovation and territorial innovation, given the fact that they originate from different fields of study. Baron further suggests to see Open Innovation in a territory. In doing so, the call from this scholar responds to our wish to consider Open Innovation through new lenses and across disciplinary boundaries.

Finally and from a policy perspective, Bror Salmelin discusses the background thinking supporting the emergence of living labs, and their rationales and their growth through the emergence of the ENoLL movement. He argues that living labs now have a strong foothold in all European regions, and have become an intrinsic component of the regional innovation systems. Furthermore, he develops the progressive convergence of and synergies between living labs and open innovation processes, taking the form of Open Innovation Ecosystems. Salmelin further develops the Open Innovation 2.0 approach which stresses the importance of the Quadruple Helix model, where the industry, academia, government and civil society collaborate seamlessly. This chapter provides insights on different instruments, and argues about the need to turn the management buzzwords into practice, relying on the existing fundamentals already present in Europe yet putting them at work, altogether, so as to build the single biggest market in the world.

## Part 2: Wisdom of the Crowd and the Role of Communities

This part starts by reviewing the literature on Open Innovation in services, taking into consideration their peculiarities, and then focuses on a specific yet critical and dominant industry, the financial services. Both reviews pave the way for further research in the field of Open Innovation in services, an area which remains largely overlooked despite the huge potential of services themselves, as well as their heterogeneity, which raises numerous challenges for research and practice.

Services have long been disregarded in (open) innovation studies. Based on this observation, Tor Helge Aas and Per Egil Pedersen discuss the peculiarities of services and to what extent those specific features may affect the feasibility of adopting Open Innovation practices in services industries. Using the framework of Dahlander and Gann[4], they perform an in-depth review of successful open innovation cases reported in the literature. Their findings unveil that the implementation of open innovation is more likely to be successful when it concentrates on the inbound side and on the non-pecuniary approach rather than on the outbound and pecuniary perspectives.

In their contribution, Patrick Schueffel and Iustin Vadana provide the first identifiable literature review on Open Innovation in an industry of paramount importance, not only *per se* but also for the effect it has on the functioning of the entire economy, namely the financial services industry. Their core focus is to explore the level of adoption of Open Innovation practices in the banking, wealth management and insurance industry, as reported in theoretical and empirical papers, and discuss Open Innovation has a potential to be more broadly applied in the industry. Their findings suggest that several organisational factors represent major obstacles for financial service firms to apply Open Innovation practices, and that some monetary constraints also refrain the adoption of a high level of openness. Nevertheless, they advocate for a wider dissemination of those practices, given the potential benefits that financial firms could reap from exploiting them fully. Furthermore, they define and characterise innovation in this industry, debate about the natural and logical openness nature of the innovation process as well as the need for structured vs. unstructured product and service development processes. Finally, their review paves the way for further research in the field of Open Innovation in financial services, which is certainly an area deserving more attention in the academic sphere as called for by Mention and Torkkeli.[5]

We continue our journey with the co-creation process involving the customers and two academic, empirical contributions. First, Maria Antikainen and Marketta Niemelä explore the role of intrinsic and extrinsic motivation of consumers in the co-creation process. Based on an in-depth analysis of two co-creation cases, the scholars elaborate a rewarding strategy which astutely combines monetary and non-monetary incentives. This strategy revolves around seven cornerstones: logic, transparency, democracy, equity, flexibility, customisable, and maintainers' active participation. Arguing that rewarding users should be an integral component of an online co-creation process, the authors point out that many online innovation communities seem to be currently lacking a solid rewarding strategy, which may lower the overall performance of the co-creation process with consumers.

Second, Jaakko Paasi's contribution aims at increasing the understanding of consumers' involvement in the new service ideation and development process. With an exploration of three cases exemplifying different forms of consumers' involvement as well as different forms of openness, which is characterised by the readability, usability and modifiability of the innovation work and of its outcome, the scholar develops recommendations on how to plan and lead open innovation activities with customers in business-to-consumers markets.

Crowdsourcing, a practice reflecting the implementation of Open Innovation, has attracted increasing interest from academics and practitioners alike. Yet much remains to be investigated about this practice so as to introduce it in an effective, efficient and productive way. In their academic contribution, Hannu Kärkkäinen, Jari Jussila and Jani Multasuo explore the use, opportunities and benefits of exploiting crowdsourcing platforms in the industrial B2B markets for the development of new products. Their findings unveil the complementarity between crowdsourcing platforms and the overall toolset mobilised in the innovation management process. The scholars further uncover that crowdsourcing platforms are of particular relevance in three phases of the innovation process: the fuzzy front end, the product development and the commercialisation stage. Using a netnographic research approach, their research concentrates on several crowdsourcing platforms used in the B2B sector and covering several types of crowdsourcing approaches, namely innovation marketplace, innovation competition, innovation event and innovation community.

In their joint academic–practitioner contribution, Pia Erkinheimo, Hannu Kärkkäinen and Jari Jussila explore in detail how crowdsourcing can influence positively the leanness and agility of start-ups' business functions. They exemplify this statement with several illustrations of pioneering companies which have used crowdsourcing platforms and to what extent, the adoption of crowdsourcing has acted as a source of value creation for these start-ups. The advantages of crowdsourcing in terms of fast execution of experiments and hypotheses testing, relying on a limited pool of internal resources are detailed, paving the way for managerial guidance and implications on how to innovate in a context of resource scarcity.

We conclude this part on the wisdom of the crowd by enlarging it so as to embrace university–industry cooperation at regional level. Päivi Iskanius focuses on university–industry collaboration, explored through the lens of Open Innovation. This single case-based research provides insights on how university–industry collaboration can be intensified by leveraging a community of practice as a channel for innovation and knowledge sharing. The role of the community of practice to solve complex, cross-industrial and interdisciplinary problems is depicted as well as the benefits that can be reaped from applying this strategy to foster dialogue and cooperation among various stakeholders at regional level.

## Section 3: Homo Sapiens & Open Innovation: Individuals on the Spotlight

This section concentrates on skills, competencies, culture and diversity management, both inside the firm, when cooperating with other entities outside its boundaries and in the specific context of mergers and acquisitions. It also deals with the management of culture and the propensity for regions to adopt Open Innovation practices based on their intrinsic features, as exemplified by Peter Prud'homme van Reine's contribution. All the following contributions benefit from joint efforts between pure academics and pure practitioners or have been written by "hybrids", skilled academics with a strong taste and experience of business venturing.

Defining the individual skills and competencies to mobilise in the innovation process, as well as characterising organisational capabilities to astutely configure, combine and reconfigure resources in view of developing novelties remain a topical and highly relevant managerial challenge. Organisational ambidexterity, a central concept in organisational theory gaining increasing popularity in technology and innovation management, is conceptually debated and empirically investigated in this chapter. Joachim Hafkesbrink and Markus Schroll elaborate on the individual competencies, as well as the organisational antecedents and competencies needed to achieve simultaneously exploration and exploitation in the specific context of open innovation processes. Their contribution further develops a framework to understand the educational needs of industry to engage into the open innovation journey, thus providing food for thought for managers, educational bodies and policymakers alike.

In their multiple case study analysis aiming at unveiling skills and competences for Open Innovation, with a specific focus on intermediaries, Wil Janssen, Timber Haaker and Harry Bouwman first delineate Open Innovation initiatives according to the structure of the network (one-to-one, user involvement, network and the possible existence of mediation), and to the degree of formality of intellectual property rights. Authors then discuss the different types of competencies that are needed according to the stage of the innovation process, namely in the idea creation, conversion or diffusion phase. Based on an explorative, multiple case study research with a sample of over 15 initiatives, such as ArchiMate and the Nokia Technopolis Innovation Mill, they unveil the competencies of the intermediary in the different projects that led to product or service innovation, which were further introduced to the market or transferred to other parties belonging to the ecosystem. The cross case analysis unearths four main types of barriers, related to the innovation process, the service and customer involvement, the firm level and the peculiarities of ICT initiatives, namely the lack of standards and the lack of ubiquitous high quality networks that hamper adoption and diffusion. The need for a shared vision is also highlighted as a key success factor, as well as the involvement of the end user. Focusing on competencies of intermediaries, they advocate for differentiated competencies according to the stage of development of the project. For example, cross-industry relationships and connection to an ecosystem might be critical in the exploration and development phase, while matching market demands and needs is the crucial element in the implementation stage.

Audrey Depeige and Stavros Sindakis examine another aspect of Open Innovation, investigating the coopetition dynamics that emerge among units and firms that belong to the same organisation in post-acquisition settings. The originality of this contribution lies in the development, review and assessment of a new framework, which combines cooperative and competitive forces within organisations, while offering new grounds for theoretical developments around the concept of coopetition. Coopetitive relationships influence organisational structures, policies, as well as coordination processes between the partner firms. Finally, the research proposes theoretical as well as operational implications for the management of intra-organisational interactions in late integration phases. These are reviewed as to the strategic decisions that are required for the management of intellectual assets and of corporate risk-adjustment schemes.

Teemu Santonen discusses about the role of diversity in fostering creativity and making innovation performance thrive, as well as the possible concerns that diversity may trigger. The scholar develops a conceptual framework for managing diversity including cultural, organisational, user-driven, cross-functional and disciplinary/cross-industry dimensions, and concludes on managing diversity with the help of portfolio management strategy.

Peter Prud'homme van Reine explores the patterns in regional culture which are enablers of an Open Innovation culture and those which may ensure that the transformation process towards an Open Innovation culture thrive. Relying on the literature in the field of regional innovation systems, the author develops a framework of dilemmas which is used to assess how regions can seemingly address conflicting demands and find a dynamic balance favourable to an Open Innovation culture, adopting a contingency perspective. Differentiating the regions according to their propensity to adopt an Open Innovation culture based on their assessment allows the author to derive policy recommendations, and identify the key elements which are conducive to Open Innovation and those which are hampering its adoption.

## The Future

This is only the beginning. The next book of this series will further explore the human side of Open Innovation, encompassing skills and capabilities for Open Innovation, the role of leadership and top management in fostering and nurturing an Open Innovation culture, the organisational readiness towards Open Innovation, the personal traits of individuals involved in Open Innovation journeys, the crucial element of trust as a prerequisite for engaging into Open Innovation initiatives. Later on, in this collective journey, we aim to address managerial issues such as the intellectual property aspects, and all the peculiarities of the contractual agreements applicable in Open Innovation contexts, the framework conditions for Open Innovation to properly function and bring fruit to communities, regions, nations and the reshaping of value constellations and ecosystems. Nurturing the education of the next generation and ensuring the transmission of Open Innovation values will also be a part of the future. In the forthcoming books, we will continue with our very wide and inclusive approach, embracing many types of firms and organisations, public bodies, from large multinational companies to SMEs, from low- to high-tech industries, across all continents and beyond.

Our vision for the future of this series is depicted below. **Openness** is central, and embodies the overall philosophy of the innovation process as seen nowadays. Openness entails the ability to listen to different, even divergent, **Opinions**, so as to be receptive to other mindsets, cultures, environments and to transform these into **Opportunities**. Individuals, teams, firms, organisations, nations, societies should capture Opportunities in a meaningful, productive, efficient and effective manner so as to create value. Value creation requires the ability to achieve a perfect **Orchestration** of capabilities, both individual and collective abilities and capabilities. Such Orchestration may benefit from **Observation**, conducted by third parties, providing impartial and fair advice, or from Observation of third parties, such as competitors, suppliers, customers, and all stakeholders involved in the value constellation. **Optimisation** is the Holy Grail, and may, at least partially, rely on the technological progress which is still booming nowadays. The use of technologies, as well as the reshaping of ecosystems, requires more and more **Operability** and interoperability between firms and systems. And only **Optimism** and willingness to engage into an Open Innovation journey can lead to fruitful and mutually rewarding relationships, ensuring that innovation delivers its intrinsic mission of building a better future and achieves societal impact. And many more **O**s will be unveiled along our joint odyssey!

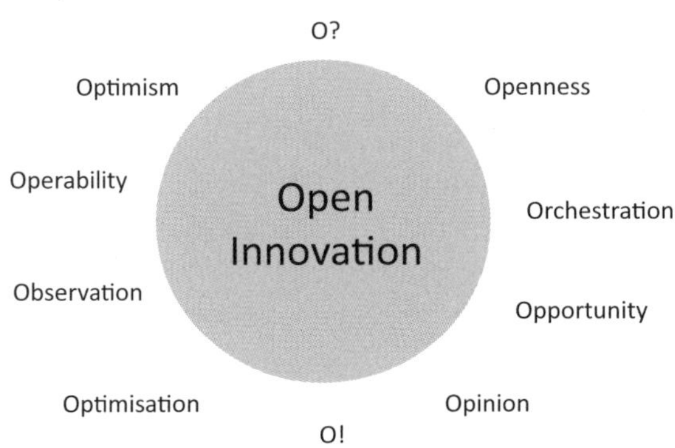

## References

1. Chesbrough, H. (2003). *Open Innovation: The New Imperative for Creating and Profiting from Technology*. Boston, MA: Harvard Business School Press.
2. Chesbrough, H. (2014). A Brief History and New Frontiers in Open Innovation, Presentation in Open Innovation Seminar held in Luxembourg May 28, 2014.
3. Dahlander, L. and Gann, D. (2010). How open is innovation? *Research Policy*, 39: 699–709.
4. Mention, A.-L. and Torkkeli, M. (2014). *Innovation in Financial Services: A Dual Ambiguity*. Cambridge Scholars Publishing, December 2014, p. 361.
5. Torkkeli, M., Hilmola Olli-Pekka, Salmi Pekka, Viskari Sari ja Käki Hannu (2008). "Avoin innovaatio Suomessa: Yritysten, korkeakoulujen ja julkisen sektorin vuorovaikutus ja yhteistyö", Tekesin katsaus 233/2008, Helsinki 2008, p. 101 (in Finnish), available online at http://www.tekes.fi/globalassets/julkaisut/avoin_innovaatio.pdf.

## Websites

Last accessed on January 21, 2015.
https://www.whitehouse.gov/open.
http://www.oi-net.eu/.
https://ec.europa.eu/digital-agenda/en/growth-jobs/open-innovation.

# Chapter 2

# OPEN QUESTIONS ABOUT OPEN INNOVATIONS

Leonid Chechurin

*Value Innovation Program Center*
*Samsung Electronics Co.*
*&*
*Lappeenranta University of Technology*
*cepreu4@gmail.com*

As an opponent who lacks extensive understanding, has not read much or done a lot of research on the subject, I have the audacity to criticise the concept of Open Innovation. I hope my engineering background, industry innovation experience and former academic footprint are enough to start this discussion. I question the novelty and the necessity of the paradigm of Open Innovation, and whether the results of research in this area have a future or even any value for practical applications.

## 1. Definitions ... and Disclaimers

It is great to have this very special chance to share thoughts and ideas on the subject of innovation. As a university professor, I have tried to study and understand what innovation is and how to support innovations. I spent several years working as concept designer in various big industries. And I worked for some time for a consulting company that supplied many Fortune 500 enterprises with new ideas and creativity support training. With all this nice and relevant-to-the-matter-of-discussion baggage, a blend of practice and theory, I am still not able to come up with solid conclusions or new findings worthy of sharing with the scientific public. However, the publications on Open Innovations that I come across from time to time bring about too many questions for me to stay silent. So, an essay seems to be the perfect way to disclaim, a chance for me to speak out, to discuss things in a relaxed manner, in other words, to speculate without shame.

To begin with, it is a good idea to check out some definitions. Surprisingly enough, the most difficult part of the term "Open Innovation" turns out to be "Open", and the main part of this chapter discusses whether it is, in fact, the proper word. Therefore, let me explain what is meant by "innovation" first.

The word "innovation" is so popular that many try it on just to be "in". Interestingly enough, in a textbook on signal processing I came across a definition of "innovation": A signal that is different from one that is expected. I think it is worth considering the generalisation of this feature: Innovation is always something wonderful ... and therefore unexpected.

Posters on the walls at many companies I have visited or worked told me that "Innovation is the new idea commercialised". This definition is very nice because it is short. And it captures an essential feature of innovation: commercialisation, or profit-making, at the end of the day. So, innovation seems to be a (sometimes very long) process of profit-making starting with a new idea. I think as far as profit-making is concerned, it is not even important if the new idea was generated or bought. I think many, if not most, great innovators started "just" with IP acquisition. But we still admire the way they made it to the market. I can refer to the case of radio: A. Popov and G. Marconi both made inventions, but only Marconi made a product, company and the first trans-Atlantic talk.

There is one disadvantage with this definition that is both understandable and forgivable for a for-profit company. Market is not the only form of life. For example, in long-term governmental or multinational projects commercialisation is hardly the driving force, while the innovative nature is obvious (space exploration, CERN, massive defence initiatives, etc.). There used to be and still exist command (not market-driven) economies that generate plenty of impressive innovations, from Sputnik to nuclear power plants. So, innovation starts with a new idea and results in a new product or technology that is used and valued by people.

Here is another popular and more general definition: "An innovation is something original, new, and important in whatever field that breaks in to a market or society".[1] However, one aspect seems to be unclear. If something new, original and important comes to life but miserably fails in the market or society, disappears shortly after the birth, should it be called innovation or not?

These and other reasons forced me to meditate on my own definition, that maybe someone will judge worth discussing: "Innovation is something new and stable." Stability can be understood here in a very brutal sense as far as material innovations are discussed. If something new comes to life and stays or works for a period, it should be called innovation. Everything that we observe around us is an innovation (or has been an innovation). There is only one way for a product or service to stay alive (be stable): To be useful, no matter whether it is in a market or command environment. Innovation can be born, amaze, but die soon and disappear if nobody buys it or supports it. In this case, we should use the term "gadget" rather than innovation.

It is interesting to apply these definitions to the world of ideas and see what happens. If a definition is good it should cover many cases. For the sake of saving room I will leave this exercise to the reader (see e.g., statistics on sharing non-material products in a country's GDP or export–import balance).

## 2. What does the Open Innovation Paradigm Bring to Science or to Innovation Practice?

Essentially, business and industry look for profits, while the mission of academia, at least historically, has always been to excavate the truth, and produce scientifically proven findings. Science and its findings are a common good, and moreover, ideas circulate openly. In addition, review by peers is one of the most critical conditions for finding truth in science.

Profit is made on truth in the perfect innovation-driven world. For example, a company's R&D reveals new material or a new method for contracting data, then launches a new product or technology based on that and wins a bigger market share. That means higher profits. If the material's formulae or algorithms are published in a scientific journal and accepted by peers, it becomes part of scientific knowledge. That means a new scientific truth. In the real business world, however, new scientifically proven knowledge, or "novel truth" is not a necessary component of profit, and I can give you at least two reasons for this. First, a new idea is only one part of innovation. Profits are influenced by its implementation, manufacturing, marketing, etc. Needless to say, good lawyers can be important for profit, or more ironically, for court fights on whose knowledge is "more novel". Second, a company does not actually need "science". Companies need research (for development) rather than science and I will insist that there is a certain difference between them. I claim that all scientific results are based on research, while not all research results become science. The difference is small but substantial, especially when we talk about openness. I believe that only results reviewed and openly published in professional proceedings deserve to be called scientific, to become the common good called science. In a sense, as far as new innovations in knowledge goes, it can become science through openness.

There are various opinions on the subject, though. Ivory tower scientists claim any scientifically proven results to be useful and able to generate profit ... let's just wait two or three hundred years. And cynical business sharks like to say that generated profit is the best proof that an idea is true.

Innovation in the world of ideas is not just something novel. The new idea somehow has to serve. Innovation must be used. Otherwise this new item becomes just an idea-gadget, a matter of fashion or short-term attraction. My point is that innovation in science or in knowledge is open, by definition, otherwise it simply does not belong to science. And science seems to be the only area where innovation means open innovation.

As far as idea generation and processing goes, sharing them, discussing them, and getting feedback on them are the healthiest ways of getting quality products. This business has to be as open as possible up to the point when it is necessary to have secrets. The words 'secrets' and 'conspiracy' are synonyms for closed (that is, concealing something), and are idea killers.

When we mean innovation, "openness" belongs to one part only, the part related to knowledge. At least this is what I learnt from the definition of open innovation by Chesbrough.[2] And this generalisation seems unnecessary, and not too ambitious to me. I mean, if we call "innovating" the part that is related to idea management, then we can say that "brushing teeth" is "body cleaning."

I have tried to experiment with other elements or stages of innovation: "open marketing", "open management", even "open profit," and I feel that it may be added to categories like "a safe without a lock," "open password" or other classical Zen Oxymorons.

Open innovation seems to have been more of an academic subject of research rather than a practical methodology or an approach to making business. The research focuses on the analysis of inflows and outflows of a company's knowledge and may generate some scientifically perfect results, the truth, or science.

## 3. What is Different in Open Innovation and Knowledge Management?

There are many things an artist can do better than a scientist. One thing that is relevant to our discussion can be described as the ability to notice, to observe, to feel a system intuitively where most people just see chaos. The artist can come up with insights, always amazing and beautiful, but often far from correct. Sometimes the beauty of the hypothesis is at the price of its correctness, but don't we willingly agree to the bargain as far as art is concerned? Leonardo's drawings of water streams document his insight in understanding the chaotic dynamic system's choreography. It took 400 years for science to come up with the concept of dynamic chaos in nonlinear systems and to reproduce the chaos at the level of knowledge. At the same time, Kondratieff's concept of long economic waves looked very beautiful. Indeed, N.D. Kondratieff in the early 20th century noticed that the world GDP is subject to small oscillations lasting for approximately 70 years. He explained these long waves by the changes in technological paradigm and it was an intriguing, beautiful hypothesis. However, it is really hard to conclude that these waves do actually exist. First, because, with just 150 years of observation data, concluding that there is a 70-year periodicity is premature. Second, because the GDP oscillations (if they exist) are negligibly small, around 5%, and can easily be generated by noise, trend elimination and other mathematical transformations required for a time series analysis. Third, a recent study shows that the theory has to be, at the least, seriously corrected to be taken as a scientifically proven truth. However, it does not hurt the beauty of Kondratieff's idea as an attempt at an interesting paradigm and it can serve as inspiration for future researchers. There are many examples of charming hypothesizes, ideas, and insights that entertain and attract the public's attention for a time period, but never become part of scientific knowledge, nor even an applicable instrument. Almost all of them are packaged in the form of impressive graphs or slogans, very general and qualitative to the extent it can be seen in everything. But, unfortunately, slogans are neither objects of science nor applicable instruments (things that come to mind are "Triple Spiral" or "Blue Ocean," etc.).

Could it be that the Open Innovation concept is still a kind of slogan rather than a phenomenon, an interesting point of view, a framework or an interpretation of the existing facts?

## 4. Is the Paradigm New or is it New Packaging of Old Things?

The analysis of financial flows in institutions has been the food for understanding and proper management of them since the beginning of the history of economy. Ability to innovate, to beat the competition with a smarter product, technology or service, has been gradually becoming more and more important over the centuries, characterised by a strengthening of market mechanisms in the economy. Also, probably starting in the middle of the 20th century, knowledge flows have become the object of analysis and research. Researchers and practitioners both have started approaching the taxonomy of knowledge, how to arrange properly knowledge generation, exchange and management. Having been packaged as a patent or know-how, knowledge has become a tangible asset. I would name the end of the 90s as the time when studies and practice somehow resulted in the Knowledge Management paradigm: Basic measurable parameters, methods, recommendations for better knowledge exchange, circulation, generation and, no doubt, protection. Economy theorists have developed approaches to evaluate the share of knowledge in a country's economy, in export–import statistics, and in a country's GDP. Of course, evaluating the amount and value of knowledge is a much harder task than evaluating material assets, and still harder than evaluating labour. It is difficult to compare ideas, evaluate the demand and supply, and, so, pricing is not routine. How, for example, to assess the value of a patent, before it becomes an object of trade? However, as with experimental physics, where smart setup helps to measure something almost unmeasurable, modern economists have come up with interesting approaches for direct or indirect evaluation of knowledge: Royalties, R&D spending, investments in personnel training, amount of new to market products, etc.

Could we think of the Open Innovation paradigm in the same way as Knowledge Management, with the emphasis on inflow/outflow knowledge flow analysis? If yes, is it so obvious that principles of knowledge management (say, IP strategies, strategies for IP licensing and acquisitions, theoretical and practical approaches to what to share and what is to be kept secret) do not work or not enough? Why, for the sake of Occam's razor, did a new entity called Open Innovation appear? What makes Open Innovation distinct? There are many examples when new successful wording has increased interest in the subject of research and brought additional investments. For example, when bifurcation theory or experiment planning were just chapters of nonlinear system theory and mathematical statistics, respectively, they were known mainly to professionals in the field. But they became really known and popular after rebranding them as "The Theory of Catastrophes" or "6 Sigma." However, it is clear to professionals in the field that these are cases of new packaging of old things. What about Open Innovation?

## 5. From Open Innovations to Free Innovations

What if innovations were really open? Can we imagine life with real open idea circulation? Since the definition of openness has already been taken, if not discredited, can we think of "free innovations?"

As writers or thought gurus say from time to time, something is really wrong and unfair in this world. But let me focus just on the part of the world related to innovations. It is true, but challenging, that small entrepreneurs can survive in the market with their inventive ideas only. They come to life with a new idea and die or reincarnate when the idea or product is not new anymore. Comparing statistics on the share of SME in GDP we can see the correlation in the innovation index from country to country. Technology- or product-driven entrepreneurship is about innovations. But the situation is different when we turn to big companies. Big business can enjoy other non-competitive benefits of size (I would mention books by Josef Schumpeter, Burton Klein and Friedrich von Hayek as witnesses to this). Indeed, a big player can use financial (and legal) resources, lobbying, alliances and acquisitions, cross-licensing, etc., etc. Invention/innovation is no longer a matter of life and death when you are big.

There is also an illusion that patenting makes all inventors equal in the face of law and justice. First of all, this is because most patents that are developed by individuals working for an enterprise belong to the enterprise, to the business owner. The idea is turned into property and becomes the subject of trade and investment for the business owner. Most inventors work for a salary; like conveyor operators, they generate ideas openly and freely, but these are open innovations within one company. The company can be good and appreciate the inventor, keeping her/him happy. But have another opinion, and the chances of an inventor to appeal for justice are small. Finally, there is little chance that even an individual patent holder can survive in a court battle with a corporation equipped with lawyers and other resources.

Big companies buy and turn individual knowledge into their own IP and after that can, in fact, enjoy preventing others from competing (or developing), thereby maximising profit, rather than developing ideas further and innovating. Now the problem can be formulated as a fundamental contradiction: knowledge is property if someone wants to make a business of it. But when it is someone's property, nobody wants to develop it. Is there a business model in which business is done without property ownership, at least as far as intellectual property is concerned?

Inspiration can come from science, where, as was already mentioned, openness is the key principle. There are papers and books — no patents, no licenses — and everyone can use knowledge for free and develop the knowledge further, only references are required. But the technology for finding new knowledge can hardly find a place in the market.

A mechanism that is closer to the market environment, a "working prototype" for the Real Openness Business Model could possibly be the Free Software Concept ("Open Source" is something slightly different). An excellent example of free software is the Linux Operation System that successfully competes with proprietary products. To make a long story short, when somebody uses your free tool, he/she is not allowed to sell the derivatives/improvements, they are to be free and open too ("viral license"). But it is ok to use the free tool as a part of your commercial product (like the OS for a car onboard computer, for example). This simple rule protects innovation from profiteers, agents who buy and sell property in the form of knowledge. I hope I can just leave keywords for interested readers: "GNU," "GPL," "copyleft licensing," "viral license," etc. So, when it is open and free, it is constantly diffused and, more importantly, improved by the crowd. And this is what I would call real openness. When it is open, your commercial benefit comes not from selling the idea, but from your services (localisation, customisation, training, etc.), and your excellence as a pioneer. Nice examples of successful businesses based on Linux Software products: Google's dirt-cheap servers; Java and Perl and Ruby for Web 2.0 Applications; free web-server software Apache that powers nearly half of all websites in the world.

But one can still argue that this concept is applicable only to software-like products. And here comes the most interesting and promising example of IP-free hardware: Arduino. It is the name of a small company (well, initially, a popular bar nearby ...) and a successful controller, that is perfect for home-made automation, labs, robotic creativity and product prototypes in the field of robotics, measurement and automation. The programming software and drives are downloadable, but this is usual. What is unusual is that all specifications and chip layouts are downloadable. That basically means you do not actually need to buy this controller, you can order it from the closest chip manufacturer, or even print it if you have the machine. No patent, no knowhow. Only one restriction: If you copy the controller, do not call it "Arduino" (call it RuDuino, FiDuino, etc.). Due to this no-IP attitude, the Arduino controller is cheaper than any proprietary one, but of the same if not better quality since many enthusiasts download, test, use, improve it and ... "share back". You can find a lot of information on the practice of using this controller, a description of projects and a great deal of free software for "do-it-yourself" projects. But what are the benefits of Arduino, what is the business model? It turns out to have been pretty good. What the company sells is expertise (customisation, training, consulting). Indeed being the pioneers in the product and creative team, they stay a generation ahead, they are known as the best experts; most of the clones are of lower quality and simply raise interest in the original device.

Arduino is, literally, a company based on giving innovation away. This is so beautiful, the idea that you are in a business because you are smart, not greedy. You ask to compensate just your working time, the time you work on the problem, but not the time you worked on the problem, not the copyright. It resembles the approach of some musicians who upload their music for free download, making their money only from live concerts.

There is a growing technology base for the Free Hardware model as well, especially cheap or free tools for chipmaking, 3-D modelling (from virtual design to 3D printing), web-based collaboration (crowdsourcing). I believe items 1 and 2 should be the technological part of equipment to be freely accessed from technology-driven innovation incubator residents (in addition to offices, internet, security, accounting support, etc.).

Is there any place for big business in the free innovation paradigm? I think there are at least three. First, when getting a solution is much more important than preventing competitors from benefitting from the results (example: Sun Microsystems and IBM paid their programmers to contribute Linux). Second, suddenly because your product can be improved by others, for free. Third, because even a big company can participate in a large free hardware project by selling the expertise, and not claiming their IP.

There are many interesting questions related to the concept of Free Innovations. Is there a trend of "going free" in modern economic life? Is there a model of business life that is beneficial for all agents (excluding patent trolls and IP profiteers)? What are the consequences of coming closer to the free innovation model (e.g., by reducing the patent term), can we simulate it? I think these are challenging topics for research in innovation management and economics.

## 6. A Word on Practical Experience

My experience says that the Open Innovation concept, as a policy or managerial strategy, has been either ignored by big business or kept as a slogan on the façade. The slogan says that we love Open Innovation, but actually means one-way knowledge flow: Your ideas and solutions are extremely welcome here for our imminent fight for market share and profits, come in and let's discuss what they are and what you want for that. It is much safer for the company to buy an "off-the-shelf" solution, to acquire the whole technology or product than rely on "closed" company research and development.

Innovation management in big business is the art of arranging and integrating existing things into a product required by the market. That was exactly what Steve Jobs did, for example, and he was hardly advised by an Open Innovation study. A company spends on R&D anyway since there are many problems left — from issues of integration to core technology development. But I have never come across a company that intentionally let knowledge leave its R&D buildings. I have never come across cases when a company gave away profits unless it was to a charity or for promo campaigns. Cross-licensing, when companies agree to share some technologies they have with other companies, even competitors, can take place during technology-intensive product development. But this is just a rational procedure to avoid "veto clinch" — when all parties equally contribute and equally benefit from the product development, a knowledge barter. Finally, companies may licenses. My experience says that buying or selling a license is not an option when the competitor is meant. I also think that selling a piece of knowledge that is not being used can happen, but it is like selling assets that are not currently in use. I do not think this can be considered a primary business for real product-oriented companies. I simply cannot imagine a semiconductor giant selling offices, for example. Finally, there are companies, in video decoding, for example, where the primary business is buying and selling patents: Patent trolls or patent profiteers. Well, according to Chesbrough's definition they are 100% Open Innovation, their main business is knowledge inflow–outflow management. But something tells me that innovations are not here at all.

In other words, when a company can solve a problem — it solves the problem. When a company cannot solve a problem — it says "open innovation is a great thing!"

Therefore, even if I can digest the term "open" as relying on, integrating and acquiring knowledge from outside, the knowledge flow goes in one direction only: into big industries. So, if the current only flows in one direction, it is a diode, a semi-conductor, not a conductor at all. How about replacing "Open Innovation" with the "Diode Innovation" Concept?

## 7. Conclusions

As stated in the introduction, this text is supposed to be an interesting reading and invitation for discussion. It is by no means a scientific contribution. Therefore, I should skip this conclusion as it might look like a scientific paper and mislead the reader. However, something is physiologically needed to mark the end of the monologue, so I'll just call it conclusions — but think "Summary".

First, I think the necessity of the term "Open Innovation" needs more proof, its novelty and fruitfulness is not evident. It sounds like a successful slogan to me, but it is not yet a scientific finding nor a promising paradigm for research. Second, my experience and basic understanding of the driving forces in business contradicts the expectation that the concept of openness can be widely used in practical applications, as a strategic management paradigm or in corporate culture. Open Innovations can be a case of non-competing cooperation, but ... my experience of cooperation has always been driven by the market and therefore competing. Finally, I believe there is still a place for real Open Innovations if we go radical and think of a Free Innovation World. In this world, a good idea is not someone's property and therefore can be freely shared, used and developed further while inventors are paid for inventing and not for inventions. So, let's see if this article is being published in an open access source ... where open means free.

## Acknowledgement

I would like to express my gratitude to Deborah Abramov, who professionally cured the text from my deep Russian accent.

## References

1. Frankelius, P. (2009). Questioning two myths in innovation literature. *Journal of High Technology Management Research*, 20(1): 40–51.
2. Chesbrough, H. (2003). *Open Innovation: The New Imperative for Creating and Profiting from Technology* (Boston, MA: Harvard Business School Press).

# Chapter 3

# OPEN DATA FOR OPEN SCIENCE: ASPIRATIONS, REALITIES, CHALLENGES AND OPPORTUNITIES

Vera Lipton

*Thomas More Law School, Australian Catholic University*
*vera.lipton@bigpond.com*

*"Imagine a world in which every single human being can freely share in the sum of all knowledge."*

Wikimedia Foundation

This chapter examines the emergent approaches to online sharing of the data underpinning scientific publications. Such data is referred to as open scientific data. Four key drivers are identified: (i) Changing societal approaches and increasing expectations that the outcomes of publicly-funded research be freely available; (ii) the emergence of data journals; (iii) changing policies of research funders and publishers; and (iv) the requirements for research organisations and individual researchers to comply with these policies and introduce research data management plans. In this chapter, the European Organisation for Nuclear Research (CERN) is used as a case study to show how open scientific data can be successfully managed and used, in innovative ways, to deliver open education in particle physics. Finally, this chapter identifies three challenges that require attention of research funders and scientific organisations. First, the challenge to devise a minimum standard for open scientific data. Second, the challenge to develop incentives for researchers and librarians, and third, the challenge to reconceptualise the ownership of open scientific data. This chapter argues that addressing these challenges can advance the preservation and sharing of open scientific data for the global public good.

*Keywords*: Open Science; Open Data; Research Data Management; Open Access to Knowledge; Open Research; e-science; Digital Science; The European Organisation for Nuclear Research.

## 1. Introduction

Scientists helped to develop Internet and communication technologies. Scientists are now using these technologies to revolutionise science in the ways it is created, communicated and disseminated in online spaces. Indeed, digital technologies have enabled researchers from a variety of disciplines and from several places to work together, in real time, to create knowledge faster and disseminate it sooner. Such improvements to scientific communications have an enormous potential to increase the value science delivers to society. Sharing of scientific data is the prerequisite for realising this potential. Good data enables good science, and digital technologies have provided the means for acquiring, transmitting, storing, analysing and re-using massive volumes of data. Research organisations and researchers are now embracing these technologies to extend the frontiers of science. This is open innovation in science, or open science.

Implementing open science requires all stakeholders — scientists, governments, research funders, publishers, lawyers and librarians — to work together. So what is the promise of open science and why should we invest resources into it? What are the necessary steps required to make it happen? What are the principles underpinning open scientific data? And what are the best ways to facilitate access to scientific knowledge while maintaining the rigour of scientific enquiry? Are there reasonable mechanisms in place to motivate researchers to release the data underpinning scientific projects? What lessons have we learnt from implementing open access to scientific publications? Can these lessons help improve access to scientific data?

This chapter answers some of these questions. The author first outlines the case for open science and then examines the emergent approaches, principles and policies driving the preparation, curation, release and sharing of open scientific data.

## 2. The Emergence of Open Scientific Data

In 1984, the United States Government started planning for a grand scientific project looking to map and decipher the entire human genome. But instead of doing it in secret laboratories on their own, this project brought together genome sequencing institutions from around the world. In the early 1980s, scientists in many countries were starting to apply computing solutions to genetics and DNA sequences, developing methods that would make it feasible to generate such data in digital formats. Encouraged by these early experiments, the Human Genome Project got underway in 1990 and was initially funded by the US Department of Energy and the National Institutes of Health. Their laboratories were joined by over 20 collaborating institutions from across the globe, including from the United Kingdom, France, Germany, Japan and China. In 2003, the International Human Genome Sequencing Consortium announced that its project was complete — 2 years ahead of schedule, under budget and with 99.99% accuracy.[a]

The success of Human Genome Project resulted from the convergence of science, technology and society in recording one whole human DNA sequence, some 3 billion letters of genetic code. This is the code that opened doors to improve understanding of human health including the detection and diagnosis of many diseases. An important accelerator was a meeting of representatives from

[a] A brief history of the Human Genome Project, http://www.genome.gov/12011239.

sequencing centres around the world in Bermuda in 1996. At that meeting scientists committed to make genomic data publicly available prior to publishing their findings in a scientific journal. The principle of free release and data sharing has been among the major achievements of the Human Genome Project and has, arguably, proved as influential as the sequencing outputs themselves. Over the years sharing of genomic data has become a more established practice and biological research has exploded. The practice of data sharing demonstrated the enormous capacity of the research community to mobilise — a shift in how scientists work together as a global community to create knowledge for the global public good.

The commitment to data sharing resulted from a fierce battle over the nature and ownership, and ultimately control, of the human genome. Two years after the Bermuda meeting a private gene sequencing company called Celera Genomics was set up in California. Celera owned a sizeable number of genome sequencing machines and aimed to create its own database of human genomic data, which would only be available to subscribers. Celera also intended to claim ownership of 300 clinically important genes and, at some stage, filed over 6000 patent applications to this end. The emergence of this powerful competitor created a fresh impetus for the Human Genome Project. One of the key UK scientists in the field later suggested that "it has not been a race but a battle to ensure that the tools to speed biomedical research were available to all."[1]

The battle went on for about 3 years. On 26 June 2000, the White House hosted a press conference that changed the rules. In front of representatives of the International Human Genome Consortium and Celera Genomics, President Bill Clinton announced that both public and private research teams were committed to publishing their genomic data simultaneously, for the benefit of researchers in every corner of the globe.[b] Later that

[b]Remarks made by the President, Prime Minister Tony Blair of England (via satellite), Dr. Francis Collins, Director of the National Human Genome Research Institute, and Dr. Craig Venter, President and Chief Scientific Officer, Celera Genomics Corporation, on the Completion of the First Survey of the Entire Human Genome Project. <http://www.genome.gov/10001356>.

year, the Human Genome Sequencing Consortium published in *Nature* while Celera's findings appeared in *Science*. The methodology presented by Celera was later criticised by many scientists, who argued that their assembly of the genome would not have been possible at the time without the data released by the Human Genome Consortium. In retrospect, Celera may have well been the first commercial user of open genomic data published in GenBank — a distributed database that stores the DNA sequence in various locations around the world. Today, GenBank supports a multimillion dollar genomics research industry.[c]

The success of GenBank has drawn the attention of governments, funding agencies and scientific organisations. The year 2003 marked 50 years since Watson and Crick's discovery of the DNA double helix. The Human Genome Project was completed in 2003. The same year marked adoption of the Berlin Declaration on Open Access to Knowledge in the Science and Humanities which called for open access to original scientific research results, raw data and metadata, source materials and digital representations of pictorial, graphical and scholarly multimedia materials.[d] The open science story continued to unfold. It is a story enabling free sharing of scientific outcomes, whether data or publications, that anyone can use, re-use and distribute without seeking the permission of anyone else.[e] Such permission may not be required because the resource is not subject to any legal control — it is, in other words, in the public domain. Or the permission may not be required because it has already been granted. In either case, to use and build upon this resource requires nothing more than access to the resource itself.[2] The resulting open scientific content and lively

[c]See for example, NIH, industry and non-profits join forces to speed validation of disease targets. <http://www.nih.gov/news/health/feb2014/od-04.htm>.

[d]*The Berlin Declaration on Open Access to Knowledge in the Sciences and Humanities* adopted on 22 October 2003. <http://openaccess.mpg.de/Berlin-Declaration>.

[e]In this sense science commons is defined as open data. See also the definition of open data by the Open Knowledge Foundation. < https://okfn.org/opendata/>.

online communities are shaping the fundamental processes of science creation, dissemination, use and teaching. These processes are taking place alongside, and are intimately connected with the evolution of digital technologies and interactive communications. Open science is developing and building upon the body of digital knowledge, data and infrastructure that it inherently generates.

## 3. The Case for Openness in Science

### *Solve great problems facing humanity*

Open scientific data is important because the need to share scientific outcomes has perhaps never been greater. As nearly every region feels the effects of climate change, conflict and food insecurity are rising and the demand for natural resources is increasing, the world looks up to science for solutions. This world is interconnected: Over 40% of population was able to access the Internet in 2013 and the number of Internet users is growing exponentially.[3] In this global digital village, open science offers hope. Hope for the those living in prosperous societies and hope for the remaining half of the globe, over 3 billion people, who live on less than $2.50 a day.[f] If anything is to secure the future and well-being of our civilisation, it is likely to be science. Yet it may not be science alone, but rather the knowledge that it imparts and the learning that it yields when it is being shared broadly and applied wisely. If science is to deliver its full value to society, it must be easily and freely accessible.[4]

### *Increase the dissemination and impact of research*

At present, the majority of science is not accessible easily and only a fraction of it is accessible freely despite the fact that scientific knowledge is plentiful and growing rapidly — doubling, on average, every

[f] See Global Issues: Poverty Facts and Stats. < http://www.globalissues.org/article/26/poverty-facts-and-stats>.

15 years.[g] Indeed, the current system of science generates massive volumes of knowledge and data. Yet much of the knowledge stays locked in institutional repositories, costly scientific journals or patent applications. Locking up knowledge does not generate the desired impact of science. Statistics confirm this: About 90% of scientific publications are *never cited* and up to half of the world's scientific papers are *never read* by anyone other than their authors, referees or editors;[5] 98.5% of patents are *never asserted.*[6] Many scientific outcomes are lost because of the failure to make them available to those who could use them and add value. This gap between the capacity for science creation and its dissemination is a 'dual tragedy' — a tragedy of science and a tragedy of society, as Australian Science Commentator Julian Cribb put it.[7]

Open science can help bridge this gap. The Internet, web and social networking have created new opportunities to disseminate scientific research, by sharing research data sooner and more widely. Much science is publicly-funded and society increasingly expects that the outcomes of public research be freely available. In the UK, Australia and many other countries, universities are the principal recipients of government funding for research. In recent years, both the UK and Australian Governments have taken considerable steps to develop mechanisms to increase the economic, social and environmental impact of science. Releasing research data is a logical step. The 2009 study of the economic effects of open access to Australian public research found that a one-off increase in accessibility to public sector R&D would produce a return to the national economy of AUD9 billion over 20 years.[h] The economic benefits of open research

---

[g]Larsen, P. O. and von Ins, M. (2010). The Rate of Growth in Scientific Publication and the Decline in Coverage Provided by Science Citation Index. *Scientometrics,* September 84(3), 575–603. The rate of doubling of the body of scientific knowledge was calculated as an average number of scientific records included in the following databases: Web of Science (owned by Thomson Reuters), Scopus (owned by LexisNexis) and Google Scholar. Duplicate entries were removed.

[h]Houghton, J. and Sheehan, P. (2009). Estimating the potential impacts of open access to research findings. *Economic Analysis and Policy*, 29(1): 127–142. Public sector R&D was defined as 'the proportion of R&D stock available to firms that will use it' and 'the proportion of R&D stock that generates useful knowledge'.

data are immense, indeed. Increased research impact is a major gain realised from investing in curating and disseminating research data. Early findings support this view. When researchers make their data accessible along with publications, they can expect an increase of up to 69% in the number of citations.[8]

## *Reduce duplication of research effort*

Open scientific data has the potential for significant savings to be realised through better targeting of scientific effort and reducing the duplication of research. Scientists, especially early career scientists, devote a great deal of their time to data collection. Moreover, the cost of collecting data for multiple research project can be high, particularly for clinical trials and drug testing.[9] If these projects complement or build on one another, why would it be necessary to provide funding for a research team to generate new datasets when another dataset already created could be used to shed light on the problem? Then, is it really necessary to create a dataset that would be used just by one research team for a single project and then be discarded? If research data can be shared, re-used and recycled, savings can be achieved. It is important that both scientists and research funders recognise this. Given the wealth of information collected in clinical trials, it is apparent that there is a variety of secondary uses that could enhance scientific advances in ways not foreseen by original authors. Indeed, the ability to access and re-use existing research can enable follow-on research and discoveries faster, cheaper, and can also facilitate the reproducibility of results.

## *Enhanced quality of scientific outcomes and methods*

Openness has been the core principle of scientific enquiry since the early days of modern science. Henry Oldenburg, German theologian and the first Secretary of the Royal Society, pioneered the peer review of scientific publications. In 1655, he referred to the printing press as:

> *"the most proper way to gratify those [who]... delight in the advancement of Learning and profitable Discoveries [and who are] invited and encouraged to search, try, and find out new things, impart their knowledge to one another, and contribute what they can to for... the Universal Good of Mankind."*[10]

Oldenburg's contemporary, Irish Scientist Robert Boyle, set two other precedents that shaped the future of science. Boyle published his results in lively English, making them accessible to those who did not speak Latin or were not trained as scientists. He also described his experiments in great detail so that others could reproduce them. In short, Boyle believed that science belonged to everyone, and the principles of science could be tested and repeated by anyone.[i]

Open science enables scientists and the general public to access and scrutinise scientific results, to 'search for the truth', as double Nobel Laureate Linus Pauling famously defined science. And the truth is often interpreted to be evidence. Science is based on the best evidence we have at the time. Evidence identifies what is true and what can be trusted. As science develops, new evidence confirms or rebuts previous evidence, resulting in self-correction. But often circumstances do not allow scientists to be 100% certain that their findings are true. So they work with the best evidence available. This evidence needs to be assessed as to its degree of reliability, which then determines the degree of confidence that can be placed on the conclusion. In these borderline cases, computer algorithms and replicated computer analyses can be used to probe the results. More often than not computers can do science faster, more accurately than humans, and increasingly, perform computations that humans cannot do. Open scientific data can serve as the springboard for computational science. Such science already brings high integration

---

[i]See for example Man of Science, Man of God: Robert Boyle. <http://www.icr.org/article/science-man-god-robert-boyle/>.

of modelling and simulations into the methodologies in particle physics, bioinformatics, earth, geospatial and social sciences.

## 4. Guiding Principles for Open Scientific Data

On 22 May 2012, at the University of North Texas, a group of librarians, scholars, researchers and university administrators gathered to discuss best practices and emerging trends in research data management. Resulting from this discussion was a vision for openness in research data titled *The Denton Declaration: An Open Data Manifesto*. The declaration includes 6 declarations, 13 principles and 7 intentions. The Principles (listed below) address the benefits of open access to data, as well as the general framework that would guide open data in science.

1. *Open access to research data benefits society, and facilitates decision making for public policy.*
2. *Publicly available research data helps promote a more cost-effective and efficient research environment by reducing redundancy of efforts.*
3. *Access to research data ensures transparency in the deployment of public funds for research and helps safeguard public goodwill toward research.*
4. *Open access to research data facilitates validation of research results, allows data to be improved by identifying errors, and enables the reuse and analysis of legacy data using new techniques developed through advances and changing perceptions.*
5. *Funding entities should support reliable long-term access to research data as a component of research grants due to the benefits that accrue from the availability of research data.*
6. *Data preservation should involve sufficient identifying characteristics and descriptive information so that others besides the data producer can use and analyse the data.*
7. *Data should be made available in a timely manner; neither too soon to ensure that researchers benefit from their labour, nor too late to allow for verification of the results.*

8. *A reasonable plan for the disposition of research data should be established as part of data management planning, rather than arbitrarily claiming the need for preservation in perpetuity.*
9. *Open access to research data should be a central goal of the lifecycle approach to data management, with consideration given at each stage of the data lifecycle to what metadata, data architecture, and infrastructure will be necessary to support data discoverability, accessibility, and long-term stewardship.*
10. *The costs of cyberinfrastructure should be distributed among the stakeholders — including researchers, agencies, and institutions — in a way that supports a long-term strategy for research data acquisition, collection, preservation, and access.*
11. *The academy should adapt existing frameworks for tenure and promotion, and merit-based incentives to account for alternative forms of publication and research output including data papers, public datasets, and digital products. Value inheres in data as a stand-alone research output.*
12. *The principles of open access should not be in conflict with the intellectual property rights of researchers, and a culture of citation and acknowledgment should be cultivated rigorously and conscientiously among all practitioners.*
13. *Open access should not compromise the confidentiality of research subjects, and will comply with principles of data security, HIPAA, FERPA*[j] *and other privacy guidelines.*

The Intentions articulate the issues of most importance at the time. These include developing a culture of openness in research, building the infrastructure that is extensible and sustainable for archiving and making the data discoverable, developing metadata standards, and recognising the intellectual property rights of researchers. By articulating the principles underpinning access to research data, the Declaration helps to fill the aspiration–usability gap as a

[j]Family Educational Rights and Privacy Act (FERPA) and Health Insurance Portability and Accountability Act (HIPAA).

mooring point for policy, administrative, and legal strategies.[k] The Principles are widely known among librarians in the United States and in other countries. In the UK, the Research Councils have articulated the common principles for open data.[l] With growing interest in the sharing of scientific data, research organisations and funders are increasingly focusing on this area.

## 5. Research Funders are the Game Changer

Over several years, leading funders of research have required grant recipients to share their data with other investigators. Yet they had no policies on how this should be accomplished. The game has changed completely in recent years, with many funders requiring the recipients of grants to enable open access to research data and, in many cases, to also submit research data management plans. Such policies ensure that data resulting from publicly-funded research are retained and can be re-used over time — usually 10 years. Research organisations and universities are largely dependent on grant funding. Suddenly, these institutions realised that in order to enable researchers to successfully compete for grants, they had to provide support in the formulation of data management plans. And libraries are doing it and researchers are changing their research data management practices as a result. Within less than 3 years, the policies introduced by research funders appear to have built a momentum for significant organisational and behavioural changes. These changes are driving the retention and sharing of research data globally.

### *United States*

The US funders of research are the leaders when it comes to open research data. The National Institutes of Health (NIH) were among

---

[k] Open Access @ UNT, 2012.

[l] See RCUK Common Principles on Data at <http://www.rcuk.ac.uk/research/datapolicy/>.

the first to introduce open access deposit of peer-reviewed journal articles in PubMed Central as a condition of receipt of grant funds.[m] The NIH also 'expects a data sharing plan for all proposals over $500,000 per year in direct costs. Some research communities have developed their own policies[n] in which sharing is expected — and executed — for all grants, not just those over the $500,000 threshold.[o]

Awareness of the need to develop data management infrastructure took a leap forward in 2010 when the National Science Foundation (NSF) announced that they would begin requiring data management plans with applications in the next grant cycle. Proposals submitted to NSF on or after 18 January 2011:

> *"must include a supplementary document of no more than two pages labelled 'Data Management Plan.' This supplementary document should describe how the proposal will conform to NSF policy on the dissemination and sharing of research results."*[p]

Importantly, the Data Management Plan is to be included with every application for NSF funding, even if the plan is a statement that "no detailed plan is needed."

The NSF policy reads as follows:

> *"Investigators are expected to share with other researchers, at no more than incremental cost and within a reasonable time, the primary data, samples, physical collections and other supporting materials created or gathered in the course of work under NSF grants. Grantees are expected to encourage and facilitate such sharing. Investigators and grantees are encouraged to share software*

[m]The NHS requires that 'an electronic version of all final peer-reviewed journal articles accepted for publication on and after 7 April 2008 be made publicly available no later than 12 months after the date of publication.'

[n]<http://www.nlm.nih.gov/NIHbmic/nih_data_sharing_policies.html>.

[o]Insel, T. Statement on Open Data, 13 June 2014, Director's Blog. <http://www.nimh.nih.gov/about/director/2013/open-data.shtml>.

[p]<http://www.nsf.gov/bfa/dias/policy/dmp.jsp>.

*and inventions created under the grant or otherwise make them or their products widely available and usable."*[q]

The US Government has taken other significant steps to enable the dissemination of scientific outcomes arising from public research. On 22 February 2013, the Office of Science and Technology Policy at the White House issued a memo *Increasing Access to the Results of Federally Funded Scientific Research*. The memo directed each Federal agency with over $100 million in annual R&D expenditure to develop plans to make "the results of unclassified research arising from public funding publicly accessible to search, retrieve and analyse and to store such results for long-term preservation."[r] The research results include peer reviewed publications, publications' metadata, and digitally formatted scientific data. The major shortcoming is that the memo does not mention metadata associated with research data. This omission is unfortunate because, in many cases, scientific data without metadata is unlikely to be re-usable.

The memo also directed agencies to ensure that intramural researchers and all extramural researchers receiving Federal grants and contracts for scientific research have data management plans in place and also mechanisms to ensure compliance with the plans. To support the implementation of data management plans, grant proposals may include appropriate costs for data management and access. Further, agencies are to promote the deposit of data in publicly accessible repositories and develop approaches for identifying and providing appropriate attribution to scientific datasets.

The memo builds on the NIH and NSF open data mandates and covers all larger federally-funded organisations. Prior to the memorandum only six federal funders of research had in place policies requiring the retention and sharing of research data: NIH, NSF, the

---

[q]See NSF Award and Administration Guide, Chapter VI — Other Post Award Requirements and Considerations, Points 4(b) and 4(c). http://www.nsf.gov/pubs/policydocs/pappguide/nsf13001/aag_6.jsp#VID4.
[r]*Ibid.*, p. 3.

National Aeronautics and Space Administration, the National Oceanic and Atmospheric Organisation, and the National Endowment for the Humanities, Office of Digital Humanities.[s]

## *United Kingdom*

The Research Councils UK (RCUK) is a partnership involving seven UK Research Councils — all major funders of research. Since the release of *RCUK Common Principles on Data Policy* in 2011, many RCUK funders have mandated data sharing with new applications. The RCUK policy on open access states:

> *"Peer reviewed research papers which result from research that is wholly or partially funded by the Research Councils:*
>
> 1. *must be published in journals which are compliant with Research Council policy on Open Access.*
> 2. *must include details of the funding that supported the research and a statement on how the underlying research materials — such as data, samples or models — can be accessed."*[t]

Unlike the United States, where institutional approaches to research data management are being developed, most Research Councils in the UK "place the responsibility on individual researchers to provide evidence that data management and sharing issues have been considered."[11] However, one of the Research Councils — the Engineering and Physical Sciences Research Council (EPSRC) — has taken a different approach. The EPSRC encourages research organisations to develop their specific approaches to data management, appropriate to their own structures and cultures. At the same time,

---

[s]Tufts University (2013). Research Guides@Tufts, *Federal Funding Agencies: Data Management and Sharing Policies.*

[t]RCUK Policy on Open Access (2012). <http://www.rcuk.ac.uk/research/openaccess/policy/>.

these approaches should be aligned with the EPSRC's expectations. To that end, EPSRC requested that Higher Educational Institutions (HEIs) develop roadmaps for open data management. EPSRC further expects that HEIs will be compliant with open data requirements by 1 May 2015. These requirements appear to have acted as a catalyst for developing data management policies and support systems in many UK research organisations.

## *Australia*

The Australian Government was among the first to invest in the development of research data infrastructure. The Australian National Data Service (ANDS) was established in 2008 to develop the Australian Research Data Commons platform[u] — an Internet-based discovery service designed to provide rich connections between data, projects, researchers and institutions. Funding was also allocated for the development of metadata tools through 'Seeding the Commons' initiative. Open research data also is a priority area for the Data to Decisions Cooperative Research Centre established in July 2014. The centre brings together researchers and industry to contribute to the development of Australia's big data capability.

The Australian data management framework, which has emerged over time, is based on four key principles:

1. The institutional data management framework is in accordance with the Australian Code for the Responsible Conduct of Research and other external legal and regulatory frameworks.
2. The research institution will support all aspects of the data lifecycle, through creation and collection, storage, manipulation, sharing and collaboration, to publishing, archiving and re-use.
3. Data management is an essential part of doing good research and supporting the research community of which each researcher is a part.

[u] <http://researchdata.ands.org.au/>.

4. Effective data management is best achieved through teamwork and collaboration between researchers, research offices, information specialists and technical support staff.

While the principles were originally drafted to outline how responsibilities between research institutions and researchers should be divided, it is now clear that open scientific data is a collective endeavour. At the same time, accountability for the preparation and curation of such data must be clearly assigned. It is for this reason that research funders, providers and researchers themselves are likely to remain the key stakeholders in this process. In this context, the Australian Code for the Responsible Conduct of Research (revised in 2007) remains the principal document guiding Australian research organisations and researchers in data management. The code states:

> *"Each institution must have a policy on the retention of materials and research data. It is important that institutions acknowledge their continuing role in the management of research material and data."*[v]

The Australian Research Council (ARC) and National Health and Medical Research Council (NHMRC) — two principal funders of national research — mandated open access to peer-reviewed publications in 2012. Starting from 2014, the Australian Research Council requires data publication for selected grants. The ARC Centre of Excellence funding agreement:

> *"... strongly encourages ... the depositing of data and any publications arising from a Project in an appropriate subject and/or institutional repository."*[w]

The NHMRC mandate does not extend to open data at this stage. Nevertheless, several Australian universities have developed

[v]The Australian Code for the Responsible Conduct of Research, Section 2.1. <http://www.nhmrc.gov.au/_files_nhmrc/publications/attachments/r39.pdf>.
[w]<http://www.arc.gov.au/ncgp/ce/ce_fundingagreement.htm>.

data management policies in recent years, with the Queensland University of Technology being the most advanced in this area.[x]

## 6. Data Journals and Publishers' Policies for Research Data Deposit

Publishers are also profoundly influencing changes in scholarly communications. Recent years have seen the emergence of data journals. The best known is perhaps Dryad[y] — a repository of research data underlying peer-reviewed articles. Dryad is governed by a consortium of scientific members that collaboratively promote data archiving, free access, reusability and citation. Membership is open to any stakeholder organisation, including journals, scientific societies, publishers, research institutions and libraries. Dryad initially covered biosciences and ecology studies and, in recent years, has expanded to other disciplines. Many libraries and research organisations now refer to Dryad as a generic data repository and recommend it for deposit in all instances where discipline specific online repositories do not exist. As a result of these practices, Dryad is increasingly becoming an interdisciplinary resource covering data from a variety of scientific fields and international sources. Data journals such as Dryad can also provide quicker access to findings, in advance of results published in paper journals or e-journals.

The growing significance of data publications has prompted established journals to expand their offerings. In early 2014, the Nature Publishing Group announced a new peer-reviewed open data publication, *Scientific Data.* The journal introduces data descriptors — a combination of traditional content and structured data to be curated in house. Such descriptors may include articles and data from multiple journals. The actual datasets will not be

---

[x] Selected data management policies of Australian research organisations are available on the Australian National Data Service website. <http://www.ands.org.au/guides/dmframework/dmf-policies-and-procedures.html>.

[y] <http://datadryad.org/>.

stored inhouse but in a recognised discipline data repository[z] or, in the absence of such repository, in a more generic data repository such as Dryad. The initial focus of *Scientific Data* is on biomedical, life and environmental sciences which appears to overlap with the initial coverage of Dryad. So, it will be interesting to see how Dryad and Scientific Data differentiate themselves and develop into the future.

Another important driver of open research data are the changing policies of traditional journal publishers who increasingly require that underlying data be made available to both peer reviewers and readers. In many cases, the publishers also specify the requirements for sufficient data description so as to facilitate re-use and validation of the research findings. For instance, the policy of *Journal of the Royal Society* (InterFace) states:

> *"To allow others to verify and build on the work published in Royal Society journals it is a condition of publication that authors make available the data and research materials supporting the results in the article. Datasets should be deposited in an appropriate, recognized repository and the associated accession number, link or digital object identifier (DOI) to the datasets must be included in the Methods section of the article. Reference(s) to datasets should also be included in the reference list of the article with DOIs (where available). Where no discipline-specific data repository exists authors should deposit their datasets in a general repository such as Dryad."*[aa]

Similarly, the journal *Nature's* policy on the availability of data and materials implies that the data should be described sufficiently to allow for validation and re-use:

> *"An inherent principle of publication is that others should be able to replicate and build upon the authors' published claims. Therefore,*

[z]Nature lists publicly-recognised data repositories on its website: <http://www.nature.com/authors/policies/availability.html>.

[aa]<http://rsif.royalsocietypublishing.org/site/misc/preparing-articles.xhtml#question15>.

*a condition of publication in a Nature journal is that authors are required to make materials, data and associated protocols promptly available to readers without undue qualifications."*[ab]

Nature also reserves the right to refuse publication to authors who fail provide evidence that they are unable to comply with the journal's requirements on data availability.

These new policies of journal publishers and funders present new challenges for researchers and research organisations. These actors have responded in different ways to the requirement of research data management. In many cases, the impetus for the development of research data management policies came from researchers applying for grants. In other cases, the changes were driven by evolving business models for data collection, curation and provision. In almost all cases, libraries have played a critical role in the formulation of data management policies, bringing their knowledge of information standards, helping to describe data and develop data citation manuals, and assisting with copyright and licensing issues. Many researchers are keen to share their data, while others are questioning whether others actually can replicate their analyses and whether the effort put into data preparation is going to bring any tangible results. Many research funders and organisations are now starting to address these concerns by recognising that a good data management policy will not only satisfy grant application requirements in the short term, but also serve as an organisational blueprint driving continuous improvement in scholarly communications for many years to come. One of such leading research institutions is the European Organisation for Nuclear Research (CERN).

## 7. Case Study: Open Data at CERN

Researchers at CERN invented the web in 1989 and the organisation is now using it to revolutionise the ways scientists develop, disseminate

[ab] <http://www.nature.com/authors/policies/availability.html>.

and communicate science, and work and learn collectively in online spaces. CERN does not have the computing and financial resources to crunch all the data it collects as part of the Large Hadron Collider (LHC) experiments in Geneva. Instead, CERN relies on grid computing powered by computer centres in many parts of the world. The Worldwide LHC Computing Grid gives a community of over 8,000 physicists near real-time access to LHC data.[ac] And portions of the data are also increasingly becoming available in the public domain as open data.

Three out of the four data detection centres at CERN — CMS, ALICE and LHCb — are committed to open data and have recently developed policies to that effect. The CERN Library is supporting these efforts by assisting with data and software curation, the development of citation practices, licensing issues, and data publication in INSPIRE — the primary information retrieval service in high energy physics. Built on CERN's Invenio Open Source Digital Library platform, INSPIRE today holds over 1 million metadata records, augmented by 300,000 full-text documents from arXiv and other sources. INSPIRE is a joint project with SLAC Accelerator Laboratory operated by Stanford University, the German Electron Synchroton (DESI) and Fermilab.[ad]

The Compact Muon Selenoid (CMS) group, involving some 3,000 scientists, released its policy data preservation, re-use and open access in 2013. The policy is developed around four levels of data complexity. In this context, the term 'data' covers:

- The content of scientific publications and data, such as figures embedded in these publications (Level 1).
- Simplified data formats for immediate re-use, for example, in education outreach activities (Level 2).
- Simulation data, along with software, workflow analyses and other documentation needed to reproduce published results (Level 3).

---

[ac] <http://home.web.cern.ch/about/computing>.

[ad] <http://inspirehep.net>.

- Raw data and software that enable full reconstruction of the CMS experiments (Level 4). Data at this level are not released in the public domain at this stage.[ae]

Level 1–3 data is released under the Creative Commons Zero waiver, which allows re-use by anyone and for any purpose. The CMS data is identified by persistent identifiers, which enable to monitor data use and help to assess the impact of the CMS program. Furthermore, it is envisaged that the open CMS data will lead to the development of a global network of users.

Early experiences with the preservation and re-use of big data (Levels 3 and 4) indicate that the biggest challenge is the loss of knowledge associated with data use and manipulation. While the LHC experiments meticulously record metadata such as event and run numbers, beam conditions and software versions used, better practice approaches to the recording of contextual metadata are still evolving. The Invenio team at CERN is working to develop templates that would enable researchers to document the steps they have taken to analyse the data.[12] CERN also actively participates in Data and Software Preservation for Open Science (DASPOS)[af] group which examines the technical issues associated with data, software and algorithm preservation and re-use.

To identify the technical tools and instructions necessary to bring the CMS open data to a wide audience, CERN ran a few pilot projects in Finnish high schools.[ag] Furthermore, the International Particle Physics Outreach Group, which began in 2005, runs master classes in high schools in over 40 countries. These master classes use data from all four data detection centres at CERN.[ah] For instance, 10% of the ATLAS data are available for students to search for a Higgs boson. These master classes also reach locations other than schools, such as

[ae] CMS data preservation, re-use and open access policy. <https://cms-docdb.cern.ch/cgi-bin/PublicDocDB/RetrieveFile?docid=6032>.
[af] <https://daspos.crc.nd.edu>.
[ag] <http://www.csc.fi/sivut/tta/national-research-data-project>.
[ah] See http://physicsmasterclasses.org/.

science centres and museums. The largest national master class program (Netzwerk Teichenwelt) runs in Germany. Every year more than 100 young facilitators, mostly masters and PhD students, take CERN data to high schools. About 4,000 students are invited to further their qualifications as part of the master class network, often choosing the topics of their research theses. Master classes offered in Greek schools are combined with virtual LHC visits where students link with a CERN physicist working on the ATLAS or CMS experiments.[ai]

## 8. Issues and Perspectives

### *Challenge 1: Minimum standard for open scientific data*

#### *Research data*

Open research data has a different meaning for different people. Research funders, researchers, librarians and lawyers all approach the term differently. Funders typically mention research data that underpins publications; researchers talk about databases and spreadsheets; while librarians are preoccupied with data, metadata, citations, and software; and lawyers would like to see 'data' described as facts, raw facts or compilations of facts in databases. This can create confusion. If researchers are to comply with the policies of funders and publishers, they need to understand what 'data' they need to make available. Similarly, if librarians are to provide effective data management services, they need to be certain about what research outputs should be considered. At present, the definitions vary not only among different stakeholders, but also among scientific disciplines.

From the perspective of researchers, clarity around open scientific data is central to both the conduct of research and preservation of its outputs. As a general guide, the Digital Curation Centre in the

---

[ai]See International Masterclasses in the LHC era, *CERN Courier*, 54(5), June 2014: 37–39.

UK recommends that researchers should consider how they will maintain access to any research data that may be necessary for enabling the validation of their research findings.[aj] One of the leading research institutions in the UK, the University of Glasgow states:

> *"research data should be interpreted as any material (digital or physical) required to underpin research. For different disciplines this may include raw data captured from instruments, derived data, documents, spreadsheets, databases, lab notebooks, visualisations, models, software, images, measurements and numbers."*[ak]

The Australian Code for the Responsible Conduct of Research provides researchers with the following guidance:

> *"...while it may not be practical to keep all the primary material (such as ore, biological material, questionnaires or recordings), durable records derived from them (such as assays, test results, transcripts, and laboratory and field notes) must be retained and accessible. The researcher must decide which data and materials should be retained, although in some cases this is determined by law, funding agency, publisher or by convention in the discipline. The central aim is that sufficient materials and data are retained to justify the outcomes of the research and to defend them if they are challenged. The potential value of the material for further research should also be considered, particularly where the research would be difficult or impossible to repeat."*[al]

In line with the guidelines, the Queensland University of Technology defines research data as:

> *"...data in the form of facts, observations, images, computer program results, recordings, measurements or experiences on which*

---

[aj] <http://www.dcc.ac.uk/resources/data-management-plans/guidance-examples>.

[ak] University of Glasgow Research Data Policy. www.gla.ac.uk/media/media_232425_en.docx.

[al] Australian Code for the Responsible Conduct of Research, Section 2.1.

> *an argument, theory, test or hypothesis, or another research output is based. Data may be numerical, descriptive, visual or tactile. It may be raw, cleaned or processed, and may be held in any format or media."*[am]

The US Office of Science and Technology Policy does not stipulate what research data is. Instead, it states what it is not:

> *"[data] ... does not include laboratory notebooks, preliminary analyses, drafts of scientific papers, plans for future research, peer review reports, communications with colleagues, or physical objects, such as laboratory specimens."*[an]

All other outputs which are accepted in the scientific community as necessary to validate research findings are included among research outputs. However, in other parts of the world, laboratory notebooks are considered as 'research data', recognising that they are necessary for reproducing research findings, especially in clinical trials. But in the US, laboratory notebooks are excluded.

## *Metadata*

Another component of open scientific data is metadata. Metadata is structured information associated with an object for purposes of discovery, description, use, management and preservation. In short, metadata is data about data; information about information. For scientific data, metadata is even more important because it provides the context needed to make sense of what would otherwise be a collection of numbers. Without metadata, any data is unlikely to be reusable. For this reason, all stakeholders involved in

---

[am] Management and Research Data Policy, University of Technology, Queensland. <http://www.mopp.qut.edu.au/D/D_02_08.jsp>.

[an] Increasing Access to the Results of Federally Funded Scientific Research. Memorandum to the Heads of Executive Departments and Agencies, p. 5. <http://www.whitehouse.gov/sites/default/files/microsites/ostp/ostp_public_access_memo_2013.pdf>.

open research data need to explicitly acknowledge that 'research data' includes metadata.

Again, there are many different standards for metadata, some of which are discipline-specific. Further clarification is necessary to arrive at a standard that would facilitate re-use of the research data and also assist libraries with targeting their efforts and resources.

### *Software*

In addition to data and metadata, the final component required for reproducibility is the code or software, or algorithms used to undertake the analyses. In a substantial number of cases, the interpretation and analysis of data is dependent on the availability of software. In some cases, sharing software code may not be permitted, as the code may be a commercial application or it can be protected by intellectual property such as patents. Yet in all those instances where the code can be shared, it should be shared and research funders need to include statements in their policies to that effect.

In summary, the initial focus of the open science data was on ensuring the release of 'data' into the public domain. Now it is necessary to provide further guidance to research organisations with regard to the definition of research data, metadata and software. While it is generally acknowledged that reproducible research requires these three components, the approach that may work best to achieve this ideal could be a minimum standard. A clear minimum standard would make open science both more realistic and possible. Such a standard would assist researchers in understanding the expectations of research funders and data publishers. The minimum standard would also enable librarians to focus their data management efforts.

## *Challenge 2: Incentives for researchers and libraries*

Preparing data for publication requires effort and time. The 2010 Survey of Machine Learning Community identified the amount of time required to document and clean up data as the major barrier to data sharing.[13] The lack of incentives for researchers to dedicate time to data preparation appears to be the nub of the issue. Unsurprisingly, compliance with policies introduced by funders, publishers and research institutions may not be the best incentive for motivating researchers, especially if their career progression remains dependent solely on the number and quality of publications and the number of teaching hours.

Libraries can assist with data preparation, and many libraries are investing significant resources into research data management. However, libraries can only assist if researchers are willing to cooperate. So the incentives need to start with researchers and measures are required. And deciding on what to measure and how to mean requires a consensus. In the context of open scientific data, the consensus is yet to be reached. The ability to evaluate the impact will increase with the growing availability of open data and the ability to count citations over time. Several methods have been suggested to track and attribute data outputs to individual researchers:

- Persistent data identifiers, such as DOI, arXiv and DataCite. Persistent identifiers help to address the problem that URL links change or fail over time.
- Linking of open scientific data with the Open Researcher and Contributor ID (ORCID) — a unique digital identifier that distinguishes one researcher from another, through integration of key research outputs such as publications, data, patents, and research grants.[ao]
- Data repository registries such as Dryad and Databib.
- Data collection registries, such as Research Data Australia. This approach seems to be favoured in Australia at this point.

In the United States, the above approaches are being implemented concurrently and it is yet to be seen what the benefits are of using one approach over another, or perhaps using multiple approaches. The key issue here will be to align metrics with the desired behaviour of researchers, so as to motivate them to release

[ao] <http://orcid.org/>.

data into the public domain. The next step is to link the metrics to performance and government funding. This process is already under way.

From 14 January 2013, the National Science Foundation:

> *"Asks a principal investigator to list his or her research 'products' rather than 'publications' in the biographical sketch section. Acceptable products must be citable and accessible including but not limited to publications, datasets, software, patents, and copyrights."*[ap]

This explicit reference to research products means that, according to NSF, scientists' worth is not dependent solely on publications. Datasets, software and other non-traditional research products will count too.[14] Scientists' worth is, of course, an indicator of their performance and impacts their career advancement. The recognition of non-traditional research outputs may encourage researchers to devote more time to these outputs.

Counting the number of research products or citations may, however, require further consideration. Citation metrics are well established for the purposes of measuring the impact of scholarly literature. In Australia, citation metrics are also being discussed as possible measures for assessing the impact of academic patents.[15] But these approaches may not be appropriate for open data, or can be appropriate to some extent only. Further discussion is needed to agree on a set of metrics that could meaningfully measure and evaluate the volume and impact of open scientific data. The research undertaken by Heather Piwowar, Founder of ImpactStory, a non-governmental organisation, is an excellent starting point for a discussion about open metrics, which she calls 'altmetrics':

> *"In the next five years ... it will become routine to track — and to value — citations to an online notebook, contributions to a software library, bookmarks to datasets from content-sharing sites such as*

---

[ap] www.nsf.gov/bfa/dias/policy/papp/papp15_1/.../nsf15_1draft.pdf.

*Pinterest and Delicious. In other words, to value a wider range metrics that suggest a research product has made a difference.*"[16]

And open scientific data is making a difference. Recent research found that publications which include well managed and curated data receive up to 69% more citations than those without data.[17] This is a powerful incentive for researchers to release their data along with publications. However, additional incentives are needed.

## Challenge 3: Data ownership requires reconceptualisation

Much of the scholarly debate about open scientific data to date has centred around the principles of intellectual property licensing, copyright and privacy laws. These bodies of law are important as they touch on the creation, preservation and use of data. But what is perhaps more important are the policies underpinning these laws. The new mandates of research funders and publishers reflect a major shift of policy. By directing and requiring openness broadly, and specifically with regard to data, the mandates dictate that research organisations take a new approach to the preservation and curation of research outcomes.

Research data is only reliable if you know where they are located, how to refer to datasets, and can assess the status of the components that make up the data, which may include elements that are subject to privacy and copyright restrictions.[18] The last bit presents challenges as a single dataset can be governed by several privacy regimes and data ownership can be difficult to establish. Many people typically contribute in the creation of an open dataset — from researchers, data collectors, and research participants, to librarians, data hosts, software providers, industry partners, and other funders. What is more, participants in research studies, such as clinical trials, increasingly would like to have a say about future use of the data. Expectations remain that some form of privacy should be the goal, along with the ability to claim ownership of research results. However, if ownership and privacy concerns block the redistribution of data on which scientific and policy conclusions are based, then those conclusions will be difficult to justify to the public who must understand them. 'We must find a balance between our ability to make and produce identifiable data, the known failure rates of de-identification systems and our need for policy and technology supported by 'good' data. If we cannot find this balance we risk a tragedy of the data commons in which the justifications for social, scientific, and political actions are available to only select a few.'[19]

With regard to ownership of research data, the property model is the wrong choice, for effectiveness and efficiency reasons. Methods for generation and sharing of scientific data increasingly fall outside the traditional ownership regime and require new approaches. In the short term, research organisations are dealing with these issues by controlling the levels of data access while still enabling re-use. In the long term, the ownership and privacy of scientific data requires reconceptualisation. Such new approaches need to encourage the growth and re-use of the scientific data commons.

## Conclusion

Open scientific data have a vast potential to improve the dissemination, quality and impact of science. Such data also promote the growth of public–private partnerships and prevent the duplication of research efforts and resources. The early stages of implementing data stewardship in open science are promising. Key players in the system — research funders, governments and publishers — have made a clear commitment to open scientific data and have created a momentum for data release. Data journals and repositories have emerged and are becoming more popular. Scientists are increasingly prepared to share data and are working with librarians to develop and implement research data management policies.

To keep the open data momentum going, all stakeholders need to work together to develop the infrastructure for long-term preservation and interlinking of data and metadata. Such infrastructure will be critical into the future as more and more data become available.

Increasing the volume of open scientific data rests on changing research culture and approaches of scientists across disciplines. Preparing data for release requires additional efforts from researchers, and, at present, these efforts are not sufficiently rewarded. Open scientific data significantly increase the number of citations publications receive, but additional incentives are required to evaluate open data. Consensus on what to measure and how is yet to be achieved. Linking the metrics to academic performance will be the next critical step.

Researchers and librarians require further guidance from research funders and policymakers with regard to the definition of open scientific data. While it is generally acknowledged that research data, metadata and software are necessary to achieve reproducibility of science, the meaning of these terms varies among stakeholders and across scientific disciplines. This chapter argues that agreeing on a minimum standard would make open science both more realistic and possible. Such a standard would assist researchers to meet the expectations of research funders and data publishers. The minimum standard would also enable librarians to focus their data management efforts, thus ultimately increasing the volume of re-usable scientific data in the public domain.

## Acknowledgements

I would like to thank to my supervisors Professor Brian Fitzgerald and Dr. John Gilchrist for their comments and support, PD Dr. Hans Peter Beck from the Atlas Collaboration at CERN and the University of Bern, and to three anonymous and exceptionally helpful reviewers.

## References

1. Rogers J. (2003). *Genome Sequencing: Wellcome News? Frontiers 03: New Writing on Cutting-Edge Science by Leading Scientists*, Tim Radford (ed.), (Trowbridge: Atlantic Press, 2003), p. 77.
2. See Lessig, L. (2006). *Code 2.0.* (Basic Books), p. 198.
3. Gasser, U., Faris. R. and Heacock, R. (2013). Internet Monitor 2013: Reflections on the Digital World. *Berman Center for Internet and Society Research Publication* No. 27, Harvard University, p. 3.

4. See Cribb, J. The Case for Open Science. *Broadcast for ABC Radio National Ockham's Razor*, November 2010 (unpublished).
5. See also Meho, L. and Yang, K. (2007). Impact of Data Source on Citation Counts and Rankings of LIS Faculty: Web of Science Versus Scopus and Google Scholar. *Journal of* the *American Society for Information Science and Technology*, 58(13): 2015–2125.
6. Lemley, M. A. and Shapiro, C. (2005). Probabilistic patents. *Journal of Economic Perspectives*, 19(2): 75–98.
7. Cribb, J. The case for open science. *Broadcast for ABC Radio National Ockham's Razor*, November 2010 (unpublished).
8. *Piwowar*, H. A., Day, R. S. and Frisma, D. B. (2007). Sharing detailed research data is associated with increased citation rate. *PLoS One*, 2(3): 308.
9. See for example Roy, A. S. A. (2012). Stifling New Cures: The True Cost of Lengthy Clinical Drug Trials, *FDA Report* (Manhattan Institute for Policy Research).
10. Oldenburg, H. (1665). *Philosophical Transactions of the Royal Society*, 1(1). <doi:10.1098/rstl.1665.0001>.
11. Davidson, J. (2014). *Mastering Digital Librarianship*, Mackenzie, A. and Martin, L. (eds.) Chapter 5, "Supporting early-career researchers" (Facet Publishing), pp. 82–102.
12. Lassila-Perini, K. G., Alverson, I., Cabrillo, A., Calderon, D., Colling, M., Hildreth, A., Huffman, T., Lampen, P., Lukka, J., Marco, T., McCauley, T. and Sonnenschein, L. (2014). 'Implementing the data preservation and open access policy at CMS.' *20th International Conference on Computing in High Energy and Nuclear Physics, IOP Publishing Journal of Physics: Conference Series* 513, p. 4.
13. Stodden, V. (2010). Reproducible Research in Computational Science: Problems and Solutions For Data and Code Sharing, ICML Workshop Machine Learning Open Source Software 2010, Haifa, Israel, 25 June 2010. <https://web.stanford.edu/~vcs/talks/VictoriaStoddenICML2010.pdf>.
14. Piwowar, H. A. (2013). Altmetrics: Value all research products. *Nature,* 493: 59. <http://www.nature.com/nature/journal/v493/n7431/full/493159a.html>.
15. IP Australia (2013). Research Performance of University Patenting in Australia, *Report Commissioned by the Department of Industry*. <http://www.ipaustralia.gov.au/about-us/news-and-media/feature-articles-listing/measuring_research_impact>.
16. Piwowar, H. A. (2013). *Altmetrics*, p. 159.
17. Piwowar, H. A., Day, R. S. and Frisma, D. B. (2007). Sharing detailed research data is associated with increased citation rate. *PLoS One,* 2(3): 308.

18. Levine, M. (2014). *Research Data Management: Practical Strategies for Information Professionals*, Roy J. (ed.), Part II. "Copyright, Open Data and the Availability–Usability Gap" (Purdue University Press), pp. 120–148.
19. Wilbanks, J. (2014). *Privacy, Big Data and the Public Good.* In Lane, J., Stodden, V., Bender, S. and Nissenbaum, H. (eds.), Chapter 11 "Portable Approaches to Informed Consent and Open Data". See also Bambauer, J. R. (2011). 'Tragedy of the data commons', *Harvard Journal of Law and Technology*, 25: 1–67.

# Chapter 4

# OPEN INNOVATION OR INNOVATION IN THE OPEN? AN EXPLORATION OF THE STRATEGY–INNOVATION LINK IN FIVE SCALE-INTENSIVE SERVICES

Karl Joachim Breunig

*Oslo and Akershus University, College of Applied Sciences*
*PB 4 St. Olavs Plass, 0130 Oslo, Norway*

Tor Helge Aas

*University of Agder, Gimlemoen 19, 4630 Kristiansand, Norway*

Katja Maria Hydle

*IRIS — International Research Institute of Stavanger*
*Forskningsparken AS Gaustadalléen 21, 0349 Oslo, Norway*

This chapter explores the strategy–innovation link in open service innovations. The increased attention to the role of the firm's external environment on innovation has important implications for strategy. However, our literature review reveals that the strategy–innovation link is ambiguously treated in extant theory, especially with respect to open- and service-innovations. Therefore, we inductively explore innovation practices in five large scale-intensive service firms to establish the link between their strategy and open innovation practices. To our surprise, we find that explicit innovation strategies lack in all five firms. The findings also suggest that the logic of innovation and strategy is hard to integrate in practice for these firms. Based on our empirical exploration, we contradict existing research on the role of strategy and open service innovations and detail implications based on our study.

*Keywords*: Open Innovation; Scale-Intensive Services; Strategy–Innovation Link; Qualitative Research.

## 1. Introduction

This chapter explores the strategy–innovation link in open service innovations. It has been argued that innovation has become more open, and this increased attention on the role of the firm's external environment on innovation has important implications for strategy.

The term open innovation was introduced by Chesbrough[1] who asserted that this approach would reduce the innovation costs as well as lead-time as firms can use external and internal ideas as well as internal and external paths to market, when aiming at advancing their technology and value proposition. Open innovation refers to "*the use of purposive inflows and outflows of knowledge to accelerate internal innovation, to expand the markets for external use of innovation*".[2]

A main characteristic of open innovations is: It implies a multiplicity of actors who act in a world of changing objectives and the elimination of the organisational boundary of the in-company and out-company origins of innovation.[3,4] The open innovation perspective explicitly postulates that its aim is to increase sustained competitiveness, and thus touches on core issues addressed by strategic management scholars. Moreover, Chesbrough[1] describes the fuzzy front end of an open innovation funnel in a way that requires strategy to prioritise innovation activities, utilise and develop resources and facilitate incremental or radical innovations across current market, new market, or other organisations market. Implicit in the open innovation conceptualisation are purposive choices on market selection, market development, resource utilisation and resource development as well as the development of value propositions — all aiming to enhance the organisational competitiveness and thus relating to the realm of strategic management. According to Teece *et al.*, the question of how firms achieve and sustain competitive advantage is the fundamental question raised in the strategic management literature.[5] In strategic management, there are many different routes to sustained competitive advantage emphasised by different authors cf. Mintzberg *et al.*[6]

We seek to bridge insight from strategic management and innovation management to address relevant aspects of an open service innovation strategy. Our literature review reveals that the strategy–innovation link is ambiguously treated in extant theory, and that open innovation entail that the firms address core strategy questions in a purposive manner. Innovations in services are assumed to be intrinsically open due to the high level of client-interaction required. However, the innovation management literature does not explain the strategy requirements an open service innovation approach entail. In fact, there are no readily available empirical studies on the role of strategy in open service innovations and the examples on open service innovation given by Chesbrough[7] are from manufacturing firms exploring the transition to offering services, and not on traditional service firms. Given the limited knowledge on the innovation–strategy link in general — and with relation to the role of strategy in service innovations with assumed high client-interaction in particular — we address the research question: *What is the role of strategy in open service innovation activities?*

This chapter is structured in four main parts. In Section 2, we assess extant innovation management literature to identify how strategy is related to open service innovation. In Section 3, we outline our research methods and explain our purposive sampling of five scale-intensive service firms. In Section 4, we present our findings, and finally we discuss these findings as well as detailing implications for practice.

## 2. Theory

Extant innovation management literature establishes that there is an important link between strategy and innovation. However, despite the identified relationship between strategy and innovation, there have been relatively few attempts in the research literature to integrate the two phenomena (with notable exception e.g., Krinsky and Jenkins.[8] Most research has either focused on innovation or strategy, and not the two phenomena simultaneously. A recent literature review[9] has assessed all articles in level 2–4 journals according to the *ABS Academic Journal Quality Guide* from 2007 up till May 2012 and identified 127 articles that address service innovation. Subsequently, we analysed these service innovation articles and identified that only 17 address strategy or strategy related topics.

The innovation management literature i.e.,[10,11] and the Product Development and Management Association (PDMA) certification work[12] explicitly state that strategy is one important dimension of service innovation management, and that an explicit innovation strategy is a prerequisite for high innovation performance. Innovation strategy refers to an articulation of the role of innovation in achieving the organisational goals[13] by aligning the overall business strategy with innovation decisions.[14] A service innovation strategy should ensure that the appropriate resources and practices necessary to develop services are present, and that the new services' characteristics, and their delivery, match customer expectations and demands.[14] Several authors[15,16] stress that it is important to set clear goals for the service innovation program as a whole, and empirical studies suggest that leading firms are likely to have an explicit service innovation strategy.[17] The PDMA innovation framework describes the product-centric development process as structured, predictable, and formal.[10]

However, open innovation and service innovation with intense client-coproduction is often described as more chaotic, unpredictable, and unstructured. According to Koelling *et al.*[18], the design of service innovation strategy depends on the connection between service interactions, service competition and complexity of service offerings. According to Menor and Roth[19], new service development strategy positions the role of development within the overall business strategy and enables management to plan for and to make available adequate resources for specific new service development efforts. A new service development strategy also ensures that the appropriate resources and practices necessary to develop services are in-line with the overall business strategy, and that the new service offering's characteristics, and its delivery, match the customer expectations and demands.[14]

None of the articles addressed open innovation specific topics such as purposive selection of markets or fuzzy front-end innovation processes. However, some articles addressed client-interaction, differentiation and resource deployment, development and change, and emphasised the difference between innovation in products and innovation in services. Moreover, Mattsson[20] focuses on customer co-creation and claims that focusing on all customers and the customer interface is better than focusing on a special type of customers (e.g., lead users). Customers are either providing the strategy or being the basis for the strategies or ideas.[21] The more emergent the service concept, the more complex the interface between the client and provider.[22] Droege *et al.*[23] and Ettlie and Rosenthal[24] elaborate on the differences between product and service innovation. Manufacturing is more likely to report the need for new strategies and structures when products are novel. Even though manufacturing and services both tend to exploit customer (externally sourced) ideas for new offerings, services are more likely to engage in shorter formal testing and to exploit general manager (internally sourced) ideas for new offering. Innovation in services thus needs another approach to strategy than innovation in manufacturing.[25,26]

The reviewed literature exposed a shortage of knowledge on the role of strategy in service innovations in general and in relation to open service innovation in particular. Nevertheless, there are issues addressed that involve strategy in the innovation process, such as the role of clients in the innovation process. However, these insights are not integrated and made explicit in relation to the strategy–innovation link. Based on the review of literature, there are in particular three areas of open service innovation strategy that needed further empirical exploration, they are as follows:

(1) **Open service innovation strategy development.** Empirical data is needed to address issues concerned with the purposive choices of market selection and development and idea generation in collaboration with external resources empirical descriptions of where strategies originate, if they are developed in interaction with external relations (in- and out-flows of ideas), and if it involves decisions on current markets, new markets, and other organisations' markets. In so doing, it is also needed to identify whether the strategies observed were predefined or if they emerge during the innovation process. Moreover, identify if types of strategies (such as innovation strategies or general business strategies) varied with respect to involvement with the external environment.

(2) **Open service innovation strategy implementation.** To enlighten the process of purposive external resource utilisation; we need empirical descriptions of how open service innovation strategy affect the service innovation practices, i.e., with respect to how ideas are prioritised, how projects are sourced with internal or external roles innovation process, how innovation resources are built to realise the strategy.

(3) **Open service innovation strategy evaluation.** To address issues of changed or improved value propositions stemming from the interaction with external resources, we need empirical descriptions of the feedback and learning processes between the firm and the external environment, identifying how the evaluation of strategy achievement occur and whether measures or control procedures give feedback to the innovation process and strategy.

## 3. Method

We deploy an explorative inductive research design justified by the lack of unambiguous theory on the strategy–innovation link for open service innovations. Qualitative data are useful for generating theory when the existing theory contains inherent contradictions or is poorly investigated.[27] The study is based on empirical case materials derived from interviews in five large international scale-intensive service firms[28] to establish the link between their strategy and open innovation practices.

The five participating firms were theoretically sampled[29–31] to enlighten issues related to strategy and open innovation. Scale-intensive services are standardised services that are produced at a large scale, mainly by large firms. Examples include bank, insurance, telecommunication, and logistics services.[28,32] This type of firms are deemed particularly relevant for this study as these firms standardise and harvest from scale advantages increasing the likelihood of observing explicit innovation strategies, and simultaneously, these firms are service firms with an assumed higher degree of client-centricity and open innovation processes.

The selected firms operates in both business to consumers and business to business markets, they all provided services both to other firms and consumers. All firms were successful in the market, as they had expanded beyond the national border to more than three countries. Thus, the firms were carefully selected to be comparable with respect to their assumed innovation performance. However, to guarantee that important variations between scale-intensive service firms were detected the five firms selected provided different types of services: Insurco and Safeco are in the insurances industry, Teleco is in the Telecommunications industry, Banco is in the financial services industry, and Mailco is a state-owned limited company within the logistics and mail services industry.

Data was collected in 25 semi-structured interviews lasting between 1 and 2 hours. Between 4 and 8 employees in each firm were interviewed. The units of analysis are service innovation projects. To

**Table 1.** Data sources

| Firms | Key informants | Management | Innovation | IT | Other | Total |
|---|---|---|---|---|---|---|
| Insurco | 1 | 1 | 1 | 1 | | 4 |
| Teleco | 1 | 2 | 2 | 2 | 1 | 8 |
| Banco | 1 | 1 | 1 | 1 | 1 | 5 |
| Mailco | 1 | 1 | | 1 | 1 | 4 |
| Safeco | 1 | 1 | 2 | | | 4 |
| Sum | 5 | 6 | 6 | 5 | 3 | 25 |

reflect both the strategy and innovation practices of the firms, informants with different roles, and from different firm levels, were chosen: Managers, business development and IT specialists (see Table 1 for details). In preparation for the interviews a questioner guide was developed based on the PDMA glossary.[33] The aim was to broadly identify innovation practices to enable the examination of the strategy–innovation link and address issues relevant to an open innovation strategy. For conducting the interviews, we followed a semi-structured interview guide, where the informants were asked open questions about how they handle innovation activities.

In order to make sense of the data, the analysis progressed in several stages and involved a blend of inductive and deductive processes.[27] First, initial coding was conducted by one of the three researchers who all took part in the interviews, and the coding was verified by one of the other members of the research team. This was repeated for each of the five firms. Then the initial codings were thoroughly discussed between the researchers as categories and findings were compared across the firms and summed up in PowerPoint format. The aim was to get a broad understanding of the innovation practices of the five firms and present initial findings to selected employees and managers in a workshop to validate the data's veracity and enhance the trustworthiness of the analysis.[34] Second, the data was coded in light of the research question, specifically examining the strategy–innovation link in each of the five firms. The coding process was partly deductive as initial categories were informed by the core dimensions identified in the literature review, partly inductive being receptive to the empirical material. Third, the findings where compared and contrasted between the firms as well as with extant theory examined in the literature review in order to extend knowledge on the role of strategy in open service innovations.

## 4. Findings

In this section, we present coded data to enlighten the three areas of open innovation — strategy development, implementation, and evaluation.

### 4.1. *Open innovation strategy development*

In all the studied firms, the strategic significance of innovation, such as time to market with new service offerings, to achieve competitive advantage is well acknowledged:

> *In our culture, we must have quick time to market, but do not have measurements or clear structure.* (Manager, Insurco)

However, innovation was primarily perceived as a tool to close the gap between the firms' current situation and their strategic ambitions.

> *The overall strategy of this bank is that they will engage with this specific customer segment. So that is the part of the strategy. But the strategy says nothing about how, or what to use ... that is only something we do ...* (Business developer, Banco)

To a large extent strategies appeared to be predefined. The findings show that the service innovations have impact on the firm, being linked to the future of the firm, and related to strategy development. However, there is little involvement of external resources in this strategy development.

> *So it is not about a top–down strategy really. We work with those who have worked more bottom–up and tried to suggest things. We get them with us, and based on that we decide on which direction we are going.* (Business developer, Insurco)

Thus, concept and idea generation are to a large extent generated internally, and the perception of the client as a resource in the innovation process is quite low. The external environment is to a limited degree utilised for new ideas. There are several informants who address that actually there are abundance of innovation ideas and initiatives and that a structured approach to prioritise between them is much more critical:

> *We actually have a surplus of ideas, but the ability to realize it, prioritize what is absolutely decisive is the key.* (Manager, Mailco)

However, all firms where very conscious about client interaction in general, and prioritised innovation activities was legitimised by current customer demand.

> *Customers expect to interact over new digital platforms — must meet the expectations of new young customers. We want to earn money and cut costs. [...] Service design is broader in relation to strategic issues and organization. New service channels, support channels will support three key result areas: Improved customer experience, increased sales, and reduced costs.* (IT Professional, Teleco)

Initially, we explored the sources of strategy to assess the role of the external environment and possible in- and outflow of ideas at the fuzzy front-end in the development of open innovation strategy. Of the five firms only Telco explicitly addressees the issue of open innovation:

> *Teleco wants to move in a more open direction to the ecosystem it sits in [...] What clearly emerged was that there was no structured process in the idea collection. So we have no process to manage*

> *ideas [...] how to drive innovation together with other players in the early stages, is something we do not have experience with. So one of my areas of focus has been how to create a structured fuzzy front end process, and to find incentives to share innovation-related knowledge.* (IT Professional, Teleco)

The findings also revealed that there are examples related to selection of markets with respect to current (both incremental and new to the market) and, however there are few examples of new markets explored and none of customers' markets. We found differences between incremental service innovation, such as automation and digitalisation, and radical new to market service innovation. Digital Postal Services (DPS), for example, is new to market service innovation. DPS is a new national digital postal system to safely send digital documents, such as insurance papers, information from the local authorities, prescriptions from MDs:

> *We started with the physical value chain of postal services, what the Postal services offer as physical post distribution. And there are a lot of similarities, distribution of documents from A to B, things to be added, but the core is similar. Core in customer segment is similar too; those that Postal Services traditionally have had, such as Energy, Telecom, public sector have a lot of documents to be distributed. So in relation to Osterwalder's business model canvas, we differentiate ourselves in relation to how we serve these customers.* (Business Developer, Mailco)

## 4.2. *Open innovation strategy implementation*

The aim of the chosen service innovation activities might be automation, customer self-service, internal organisational consequences as customers were now doing the job that internal employees would normally do. These changes however produce changes to the services delivered, to the internal organisation and knowledge sharing within the firms as well as changes to the way the firm interacts with its clients. Overall we find that service innovation is believed to take a more holistic approach to innovation than product innovations and that service innovation are more incremental:

> *Innovation is often associated with completely new products. Service Innovation implies that we look at everything that is around — business process automation and offshoring is then included — not to mention customer involvement and satisfaction.* (Manager, Insurco)

The findings show that the overall business strategy focus influence innovation activities:

> *It is clear that strategy is very important to create a plan of action. An action plan is based on the strategy.* (Business developer, Banco)

The findings indicate that the five service firms lack explicit innovation strategies but that general business strategies give direction and goals for innovation activities in general:

> *Insurco's Innovation Strategy is missing.[...] Do not aim for it — do not look at innovation as a separate exercise. Completely impossible to engage in customer-oriented development without conducting innovation — we do it but it is driven from the bottom up — not as a strategy from the top, with dedicated roles and processes it.* (IT specialist, Insurco)

An explicit innovation strategy is not considered necessary, as innovation is perceived as a tool to reach the business strategy:

> *We have a strategy, but not specifically on innovation. Innovation will support the overall strategy. [...] Innovation is a tool.* (Business developer, Teleco)

With respect to the utilisation of external resources in the innovation projects we found that innovation activities in the five scale-intensive service firms to a large extent were sourced internally, and that client involvement in the innovation process is to a large extent assumed to entail prototype testing and feedback through i.e., focus groups:

> *All redevelopment that we conduct should be tested by our customers, but it does not always happen. Can test the old channels; instead we use focus group.* (Manager, Banco)

In all the firms issues related to knowledge transfer was addressed. Not just issues related to sharing potentially business sensitive insight with external parties but also knowledge sharing within the organisation.

> *The challenge is to collect the insights we have internally. We as an organization encounter enormous information. There are a lot of reports and customer surveys. We collect a vast material, much quantitatively, and we have plenty of customer and market information.* (Business Developer, Safeco)

Issues related to knowledge transfer were addressed as related to strategy implementation both with respect to organisational structure and incentives to promote certain knowledge sharing practices.

> *This means that you need to have a kind of flexibility so you can coordinate the decision points in the two companies. And it's a very different kind of process than if you then decide speed internally.* (Manager, Teleco)

## 4.3. *Open innovation strategy evaluation*

As described above, our findings show that the problem is not the number of ideas, but the ability to connect innovation and strategy so that new ideas find their resources to develop and implement the ideas. Consequently, several informants address issues related to the innovation decision process, how to structure new ideas and that the critical issue is not ideas but the capacity to prioritise between them:

> *We have a large number of new ideas, the challenge lies in taking ideas to commercialization. It is mostly about to implement, prioritize and develop good ideas.* (Business Developer, Safeco)

The findings show that the challenge with evaluating innovation performance and the contribution of innovation on the achievement of strategy. Apart from general organisational performance there is feedback on the innovation process. To measure innovation in advance and assess the potential contribution to the strategic goals are deemed difficult. For example, customer experience is a strategic area where a large strategic gap should be closed with innovation initiatives, and yet in the innovation process described weeds away those ideas due to the inability to express innovation outcomes in financial terms:

> *In the business development department, we try to create [innovation project] mandate and measures — what is the resources, time, scope, and type of impact we believe the initiative can bring. Once that is done, we go into a creative idea phase to see what's our options — create hypotheses to see what is possible, technical, marketing and resources [...] So we establish a formal project and getting resources to manage the project.* (Business Developer, Safeco)

The informants also express that it is difficult to measure innovation outcomes in monetary terms, or to assess financial implications short-term vs. long-term:

> *It's hard. Because the prioritization requires a business case with quantifiable measures.* (Business Developer, Safeco)

Thus, the challenge is to make strategic decisions in the innovation process. It appears that innovation decisions are made more on the bases of what is currently identified as profitable rather than on a long-term strategic choice.

> *If 4-generation mobile network had been evaluated at the beginning of the innovation process with the present structured stage-gate approach it would never have been decided on.* (Manager, Teleco)

Our findings expose that service innovation activities appear to be more linked to the existing business strategy than to an innovation strategy in the five large scale-intensive service firms. Innovation is considered to be a tool to realise strategy. Client expectations are used to legitimise innovation activities, but often they have clear short term, cost reducing or efficiency enhancing goals.

The role of the client in the innovation process is much weaker than anticipated. Clients are to a limited degree utilised as a resource in the innovation process. Clients interact with innovation activities during tests of prototypes or for market research purposes. The firm's external environment is to a limited degree utilised as a source of innovation ideas. Although there are abundance of innovation ideas and initiatives, a structured approach to prioritise between them is much more critical.

Service innovations involve an external environment that appears chaotic, unpredictable, and unstructured and all the firms are making an effort to establish innovation processes that enhance structured, predictable, and formal decision making to take place. Paradoxically, the result is that many truly strategic long-term decisions do not qualify and are not initiated as innovation projects due to the structured process that requires i.e., innovation outcomes expressed in financial terms. The challenge is to make strategic decisions in the innovation process.

## 5. Discussion and Conclusions

We coded our data to inform open service innovation strategy development, open service innovation strategy implementation, and open service innovation strategy evaluation in the innovation practices studies. Open service innovation strategy development entails the purposive choices of market selection and development and idea generation in collaboration with external resources. To our surprise, we find that explicit innovation strategies lack in all five firms. Open service innovation strategy implementation entails how open service innovation strategy affect the service innovation practices, i.e., with respect to how ideas are prioritised, how projects are sourced with internal or external roles innovation process, how innovation resources are built to realise the strategy. Open service innovation strategy evaluation entails the feedback and learning processes between the firm and the external environment, identifying how the evaluation of strategy achievement occur and whether measures or control procedures give feedback to the innovation process and strategy. The findings also suggest that clients are utilised to a lesser degree in innovation activities than suggested in literature and that the logic of innovation and strategy is hard to integrate in practice for these firms. Innovation is thus considered as a tool to realise strategy. Based on the findings of this study it appears that service innovation activities are more linked to the existing business strategy than to an innovation strategy. These observations contradict extant innovation management literature, and we contribute to the literature with our five main findings:

- The distinction between business strategy and innovation strategy is less explicit in service innovation.
- The utilisation of clients in innovation activities is to a limited degree strategic.
- There is an abundance of innovation ideas that are challenging to prioritise between.
- Innovation strategy requires linking innovation resources, both internally and externally.
- There is a conflicting logic between innovation and strategy in services.

Our empirical exploration confirm that service innovations need to be distinguished from product innovations and that service innovations necessitate a broader link to business strategy due to the more game changing nature of the service innovation process. Moreover, open service innovations involve an external environment that appears chaotic, unpredictable, and unstructured and all the firms are making an effort to establish innovation processes that enhance structured, predictable, and formal decision-making. Paradoxically, the result is that many truly strategic long-term decisions do not qualify and are not initiated as innovation projects due to the structured process that requires i.e., innovation outcomes expressed in financial terms. The challenge is to make strategic decisions in the innovation process.

## References

1. Chesbrough, H. W. (2003). Open Innovation: *The New Imperative for Creating and Profiting from Technology* (Harvard Business School Press, Boston).
2. Chesbrough, H. W. (2006). *Open Business Models: How to Thrive in the New Innovation Landscape* (Harvard Business School Press, Boston, MA).
3. Fung, V. K., Fung, W. K. and Wind, Y. R. (2007). *Competing in a Flat World: Building Enterprises for a Borderless World* (Pearson Prentice Hall. Upper Saddle River, New Jersey).
4. Westerlund, M. and Leminen, S. (2011). Managing the challenges of becoming an open innovation company: Experiences from living labs. *Technology Innovation Management Review*, 1(1): 19–25.
5. Teece, D. J., Pisano, G. and Shuen, A. (1997). Dynamic capabilities and strategic management. *Strategic Management Journal*, 18(7): 509–534
6. Mintzberg, H., Ahlstrand, B. and Lampel, J. (1998). Strategy Safari (Prentice Hall. London).
7. Chesbrough, H. W. (2011). Open Service Innovation: The New Imperative for Creating and Profiting from Technology (Jossey-Bass, San Fransisco).
8. Krinsky, R. J. and Jenkins, A. C. (1997). When worlds collide: the uneasy fusion of strategy and innovation. *Strategy & Leadership* 25(4): 36–41.
9. Kurtmollaiev, S. and Pedersen, P. E. (2015). Managing Service Innovation: Yesterday, today, tomorrow. Working Paper (unpublished), Center for Center for Service Innovation (NHH — Norwegian School of Economics, Bergen, Norway).
10. Froehle, C. M. and Roth, A. V. (2007). A resource–process framework of new service development. *Production and Operations Management* 16(2): 169–188.
11. Kahn, K. B., Barczak, G. and Moss, R. (2006). Dialogue on best practices in new product development PERSPECTIVE: Establishing an NPD best practices framework. *Journal of Product Innovation Management* 23: 106–116.

12. PDMA, in NPDP Certification Study Guide (2012). Available from: http://www.pdma.org/p/cm/ld/fid=26.
13. Cooper, R. G., Edgett, S. J. and Kleinschmidt, E. J. (2001). Portfolio Management for New Products (Basic Books, New York).
14. Menor, L. J. and Roth, A. V. (2007). New service development competence in retail banking: Construct development and measurement validation. *Journal of Operations Management* 25(4): 825–846.
15. Easingwood, C. J. (1990). Service Design and Service Strategies. In 1st International Research Seminar in Service Management. Institute d'Administration des Enterprises, Aix-en-Provence, France.
16. Johne, A. and Storey, C. (1998). New service development: A review of the literature and annotated bibliography. *European Journal of Marketing* 32(3/4): 184–251.
17. Cooper, R. G., Edgett, S. J. and Kleinschmidt, E. J. (2002). Optimizing the stage-gate process: What best-practice companies doing. *Research–Technology Management* 45: 21–27.
18. Koelling, M., Neyer, A.-K. and Moeslein, K. M. (2010). Strategies towards innovative services: Findings from the German service landscape. *Service Industries Journal* 30(4): 609–620.
19. Menor, L. J. and Roth, A. V. (2008). New service development competence and performance: An empirical investigation in retail banking. *Production and Operations Management* 17(3): 267–284.
20. Mattsson, J. (2010). Developing a strategic abstraction tool for service innovation. *Journal of Strategic Marketing* 18(2): 133–144.
21. Ottenbacher, M. C. and Harrington, R. J. (2008). New service development of entrepreneurial innovations in the IT sector: Identifying levers for success. *International Journal of Entrepreneurship & Innovation* 9(1): 21–31.
22. Möller, K., Rajala, R. and Westerlund, M. (2008). Service innovation myopia? A new recipe for client–provider value creation. *California Management Review* 50(3): 31–47.
23. Droege, H., Hildebrand, D. and Forcada, M. A. H. (2009). Innovation in services: Present findings, and future pathways. *Journal of Service Management* 20(2): 131–155.
24. Ettlie, J. E. and Rosenthal, S. R. (2011). Service versus manufacturing innovation. *Journal of Product Innovation Management* 28(2): 285–299.
25. Lightfoot, H. W. and Gebauer, H. (2011). Exploring the alignment between service strategy and service innovation. *Journal of Service Management* 22(5): 664–683.
26. Nordin, F., Kindström, D., Kowalkowski, C. and Rehme, J. (2011). The risks of providing services: Differential risk effects of the service-development strategies of customisation, bundling, and range. *Journal of Service Management* 22(3): 390–408.

27. Graebner, M. E., Martin, J. A. and Roundy, P. T. (2012). Qualitative data: Cooking without a recipe. *Strategic Organization* 10(3): 276–284.
28. Pavitt, K. (1984). Sectoral patterns of technical change: Towards a taxonomy and a theory. *Research Policy* 13(6): 343–373.
29. Flyvbjerg, B. (2011). Case Study. In Denzin, N. K. and Lincoln, Y. S. (eds.) (Sage, Thousand Oaks, CA).
30. Eisenhardt, K. and Graebner, M. (2007). Theory building from cases: Opportunities and challenges. *Academy of Management Journal* 50(1): 25–32.
31. Eisenhardt, K. (1989). Building theories from case study research. *Academy of Management Review* 14(4): 532–550.
32. de Jong, J. P., Bruins, J. A., Dolfsma, W. and Meijgaard, J. (2003). In *Innovation in Service Firms Explored: What, How and Why*? EIM Business & Policy Research, Zoetermeer, Holland, EIM Business & Policy Research.
33. PDMA (2006). *Glossary for New Product Development*. Available from: http://www.pdma.org/p/cm/ld/fid=27.
34. Lincoln, Y. S. and Guba, E. G. (1985). Naturalistic Inquiry (Sage, Beverly Hills, CA).

# Chapter 5

# SUSTAINABLE INNOVATION: SOLVING WICKED PROBLEMS THROUGH INNOVATION

Antti Hautamäki

*Agora Center, University of Jyväskylä, P.O. Box 35, FI-40014*
*University of Jyväskylä, Finland*
*antti.hautamaki@kolumbus.fi*

Kaisa Oksanen

*23, FI-00023 Senior Specialist, Prime Minister's Office*
*P.O. Box Government, Finland*
*kaisa.oksanen@gmail.com*

The purpose of this chapter is to present a new concept of sustainable innovation and offer analysis and constructive ideas regarding the relevance of innovation for solving wicked problems. The drivers of innovation are changing. As a result innovation policies and processes all over the world need new approaches to tap into the undiscovered innovation potential. By reviewing the literature and examples from today's innovative practices and policies we explore the transition of the concept of innovation from technological and economically measured innovation towards sustainable innovation. The concept is defined by impact orientation and systemic and inclusive approach. Sustainable innovation means innovation that balances the long-term influences of the process and the output with the needs of people, societies, the economy and the environment. In addition, sustainable innovation democratses innovation as it aims at including all people.

*Keywords*: Sustainable Development; Sustainable Innovation; Systemic Innovation; Inclusive Innovation; Wicked Problems; Innovation Policy.

## 1. Introduction

This chapter explores how traditional paths to innovation are losing their edge in innovation competition, and presents new perspectives to innovation processes and policies. The nature of the chapter is conceptual, providing a new framework from which one can consider the current development of innovation. Because innovation is associated with problem solving, the special innovation challenges of today are related to wicked problems, those challenges in life and society that are particularly complex, multi-faceted, and that require creative approaches.

One common denominator of wicked problems is sustainable development. World Commission on Environment and Development (1987) defines sustainable development to development which meets the need of the present without compromising the ability of future generation to meet their own needs.

Sustainable development has three dimensions: economic, environmental and social.[1] Following these definitions, we label the new innovation concept as *sustainable innovation*. Through our research, we seek to uncover the various dimensions of sustainable innovation and to explore its inputs to the conceptual elaborations of innovation and in managing wicked problems.

Wicked problems are complex issues where solution requires extensive cooperation and many actors[2], but when succeeded the solutions provide means to tap into a significant, long-term innovation potential. The role of innovation in solving great challenges such as climate change or water scarcity is indeed becoming increasingly important.[3] Similarly, the business models are changing together with innovation.[4] Pioneering entrepreneurs introduce new products and services, expand the range of global knowledge networks, and most importantly, challenge established business and innovation interests with new approaches.[5] What is important for solutions is the systemic nature of wicked problems. Therefore, sustainable innovation must be holistic and avoid partial optimisation.

The nature of innovation is shifting from the application of new technology to the delivery of meanings, values and solutions.[6] Today's innovators need to develop new capabilities covering the entire innovation chain, from basic research to products, services and markets. Solving wicked problems through innovation further enhances the need for new capabilities, because innovation is not grounded in convention, but it challenges the existing mindset and ways of doing. Both innovation and wicked problems have to be dealt within a context of uncertainty and risk, and both require collective.[7]

During the last decades, wicked problems have been a hot topic especially within public administration and policy studies.[8] As public organisations, companies, NGO's and citizens are all interested in creating solutions to wicked problems, more efficient identification of problems and more collaborative solution creation are needed. Similarly, innovations are more often produced in co-creation among diverse individuals and groups, not by firms or institutions alone. The basic challenge is that the problem solvers often possess conflicting views of the problem, solution methods and the legitimacy of possible solutions.[9] The best solutions are however created if all stakeholders are able to find their place within the problem-solving network; this requires an *inclusive approach to innovation.* Ideation and discussions should happen in the shared arenas, where organisations together with opinion leaders and other central figures guide the innovation processes and value creation.[10]

In order to develop the arguments we have outlined and analysed the relevant theoretical contributions to innovation and utilised empirical examples of sustainable innovation. The research material consists of innovation literature from the academia, business and governments, and of selected examples of innovation practices that widen the scope of innovation all over the world. Using a literature review and data from today's innovative practices and policies, the chapter explores the changing concept of innovation and then continues to investigate the impact of innovation and systemic approach to innovation. This analysis lays the theoretical framework for analysing sustainable innovation. The investigated approaches underline innovation as a solution to complex problems and as embedded within a certain system to that is affected by and affects the solutions in many ways. As a result, the chapter portrays the concept of sustainable innovation and assesses its implications in solving wicked problems. Section 3 outlines the innovation policy for sustainable innovation.

## 2. The Concept of Innovation in Transition

The concept of innovation was long and limited to the relationship between technology, productivity and economic growth. Schumpeter's classical innovation concept consisted of new products to market, new production methods, new markets, new sources for raw materials or partially manufactured goods, and new industry organisations.[11] From the beginning, innovation was distinguished from ideas and inventions, since it was defined as a new useful item that had been adopted into use. Innovation was for long market-driven and the process was based on delivering new customer value in the marketplace and reducing costs for producers.[12] Applying technological innovations has indeed led to a more effective use of productive resources, and to new economic solutions such as new products, processes, and services. Furthermore, several studies demonstrate a virtuous cycle in which R&D, innovation and productivity mutually reinforce each other.[13,14]

Fundamentally, in contemporary contexts, innovation means introducing something new and the concept is not limited to technology or economic growth. Innovation intersects with a variety of sociocultural issues, not only technology, markets, and products. Today's innovation comprises of services and non-commercial social innovations.[15] In addition, innovation has been characterised by different competing modes, for example, one based on the production and use of codified scientific and technical knowledge, the so-called science–technology–innovation (STI) mode, and the other based on the doing–using–interacting (DUI) mode.[16]

To sum up, the definition of innovation has been expanded and separated from purely technical innovations. Innovation can be a new product or a new service, a new organisation, a new action model, or a new concept. In the past decade, especially the theory of open innovation has received considerable attention.[17–19] Also the extensions of innovation to service innovation and societal innovation have expanded in the recent years.[20,21] Social innovation is about developing new ideas to tackle social problems or meet social needs; it may be a new product, service, initiative, organisational model or approach to the delivery of public services. In practice, social innovations can be new services such as the NHS Direct or new social enterprises like *The Big Issue* — a magazine sold by homeless people in the UK.[22]

Although the talk about innovation has been a Western phenomenon, innovation activities are expanding in developing countries, especially in East Asia.[23] As an interesting example, India has its own unique approach to innovation called *jugaad* innovation that means overcoming wicked problems by improvising an effective solution using limited resources. In practice *jugaad* innovators create solutions with marginalised, often low income and non-traditional groups in mind, both as consumers and employees. The inclusive approach is reflected in their very low-cost products and services that make these innovations accessible to a greater number of consumers, while generating employment and helping solve pressing problems in the healthcare, energy and technology sectors.[24] The *jugaad* approach to innovation is widespread in India, and it aims not only producing solutions in India but in the whole world. Another example developed in India, the education-enterprise model of Indian Institute of Kharagpur offers pioneering views of the

Indian innovation by providing an entrepreneurial education in intended to solve various social problems and to create and sustain social values such as common community goals (see e.g., Indian Health Kiosk[25]).

These examples demonstrate a significant shift in the defining element of innovation: From maximising profits to solving wicked problems, all while having a positive effect on society as a whole. However, these new approaches often appear under-developed when people apply the traditional manufacturing logic to contemporary innovation outcomes.[26] New insights and deeper understanding of the new perspectives of innovation are needed: Perspectives based on co-creation and new values.[27]

New approaches to innovation emphasise collaboration and inclusivity. Wicked problems are similarly complex issues where solution requires extensive cooperation and many actors, and the literature on wicked problems indicates that they are very hard to solve.[28] Because of the complexity and difficulty in defining wicked problems, the solutions can be radical or disruptive innovations.[29] The emphasis of this chapter rests upon the inclusive and the solution-creating aspects of innovation, as well as on the impact of the innovation, which we argue to be in the core of new perspectives and practices of innovation. Innovation is not just an idea or invention, but it has actual impacts on the environment in which it is used, and the impact will reveal the true value of the innovation.

## 2.1. *The impact of innovation*

The innovation process is often presented in a linear format that includes different phases[30,31] or in a cyclical format where the order or phases can vary more and innovations may deviate from the general cycle.[32–34] We follow the latter conceptualisation.

Innovation is described to have its lifecycle from a concept to practice.[35,36] This lifecycle must be further analysed. In fact, there are four elements in the lifecycle of innovation: idea, invention, implementation, and impact.[37] We call this approach the 4i-model

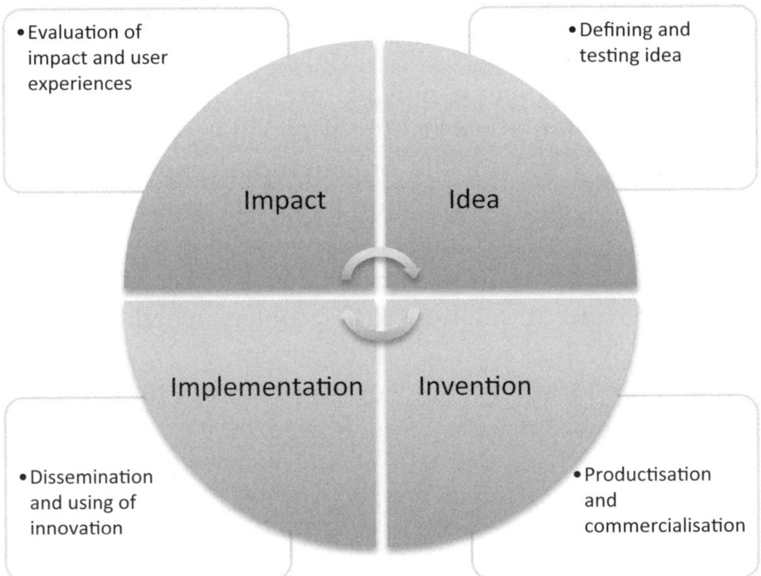


**Figure 1.** The 4i-model of innovation

of innovation (Figure 1). The innovation formula for this model is: Idea → Invention → Implementation → Impact.

The driving force of innovation process is a sustaining problem demanding a solution. The idea stage means introducing a tentative solution to the problem. It is a new concept or a new combination of existing ideas and knowledge. The invention stage involves developing modes and prototypes of the idea. Between the emergence of ideas and the implementation of ideas are the processes of qualifying and testing. In the implementation stage, production begins and the innovation is introduced to the users and customers. The focus and outcomes of the idea generation and implementation processes are quite distinct. Ideation is free and creative, while the implementation process needs to be more disciplined and purposeful. In the impact stage, the effect of innovation and the actual ways of use are actualised and assessed. Evaluations and new insights can be added in any stage and user experiences often change the process and bring about new ideas. Very often, these stages run in parallel and influence each other through feedback. User feedback also might lead to a new idea.

What keeps the 4i-circle moving, are the flow of ideas and the ongoing identification of wicked problems. In innovation practices a lot of emphasis is laid on the organisations' capability to gather ideas, to network, and to collaborate.[38,39] However, ideation and networks do not produce innovation without motivation. Wicked problems are an important motivational source because, ultimately, innovation provides a solution to a problem worth solving. This point often is overlooked when in discussion of creativity or ideas generation. Pure ideation rarely creates successful products; it takes a real, persistent problem, a genuine need that requires resolution. Real solutions also have real impacts on the environment and users.

Innovation creates new practices, and leads to changes in the structures of organisations and in the actions of people. A good example is the cell phone. The technical innovations of the cell phone are related to signal processing, to batteries or to display etc. but maybe the most important social innovation lies in its use as a communication tool between people and as access to Internet. The development of the cell phone and the creation of smart phones have impacted tremendously the whole communication culture.[40,41]

The impact stage is often ignored in the innovation research, because innovation is considered ready when it is implemented. In addition, there is a general assumption that innovations are always useful, valuable and good in nature. These kind of qualities are impossible to verify without considering the impacts of innovation. Innovation could be an economical success, but socially a disaster, because of its impact on social practices (e.g., excessive marketing of infant formula in developing countries[42]). However, the goodness of innovation has not been widely studied. Some researchers have pointed out that it is possible that innovation is harmful or uneconomical from the point of view of an individual or a social system[43,44], but the given nature of innovation needs further investigating.[45]

Innovation is often defined as an invention, which is successfully commercialised.[46] The impact is not seen similarly relevant, except perhaps in terms of revenue. Beside the economic bias, this definition also misses the most important aspect of innovation: changing the world. In economics, impacts of innovations are classified as externalities, which refers to "the direct impact of one variable on another variable that does not take place through a market transaction."[47] Externalities can be negative such as pollution or positive such as learning by interaction. To analyse the real impact of innovation we need other approach instead of only economic analyses of market relationships.

One general approach to study the impact of innovation is to apply the IOOI-model of impact. In the IOOI-model, the impact chain is divided to four phases: input–output–outcome–impact. Output means concrete results of innovation process such as tablet computers. Outcome is the effect of using the innovation such as using the new tablet. Impact is more related to the meaning of innovation when it is utilised widely and extensively throughout the system: for example, the adoption of tablets in the working life causes changes in working conditions.

In IOOI-model exact measures for impact are developed. They might be useful also in evaluating the impact of innovation. For this kind of evaluation, in addition to econometrics we need considerations from social and cultural sciences. If we take the impact to be a central element of innovation, the whole thinking about innovation must be changed. When traditional mindset mainly considers the return on investment (ROI) the new mindset means the consideration of *return on behaviour* (ROB). The point of view must change from profiting from innovation into intended impact of innovation: How to get intended change in society and environment. This seems to be the central aspect of sustainable innovation.

## 2.2. Sustainable innovation

Sustainable development challenges companies to develop products and services for new clean-tech markets that demand better control over the lifecycles of these products and services, the use of recycled materials, when possible, energy efficiency, and attention to quality of life. Nidumolu *et al.* even argue that there is no alternative for sustainable development.[48]

Sustainability as a driver of innovation means that companies must be one step ahead of the others and be able to anticipate where the market will demand sustainable development in products and services. Companies must also be able to stand clearly against the narrow interests of markets while linking their mission and practices to social, economic and environmental responsibility. Sustainable innovation therefore means that innovation should balance the long-term impact of the innovation and the actual innovative outcomes, to satisfy the needs of people, societies, the economy and the environment.[49]

Sustainable innovation has its roots in sustainable development, and it is based on ethically, socially, economically and environmentally sustainable principles. Similar principles can be seen in *eco-innovation*[50,51], in *frugal innovation and engineering*[52], in *jugaad innovation*[53], and in the rise of *shared-value* mindset[54], but the wider concept of sustainable innovation needs to be elaborated thoroughly.

This is highlighted in the mindset of *jugaad* innovation: When other resources are scarce, the creativity of an entrepreneur and innovator is the key to success. A well-known example is the Tata Nano Project, in which *jugaad* innovation was employed in a complementary fashion at various stages of the process. The Tata Car demonstrates inclusiveness in terms of drawing on diverse talents to control the cost of designing, manufacturing, R&D practices, and so forth.

In business, the motivation for innovation has been to create superior competitiveness in the market place. Traditionally, this has been accomplished through two basic strategies: To cut costs or to create better products as compared to one's competitors (cost leadership or differentiation strategies[55]). Sustainable innovation, however, offers a third competitive strategy: To create products or processes with market-desirable features, such as durability, locality, or material or energy efficiency. The motivation for sustainable innovation lies in combining competitiveness, the well-being of people, and sustainable solutions. Emphasising well-being as a strategy means not only addressing specific needs of people but also creating future-oriented and sustainable solutions. Since consumers are demanding sustainable products and services and are willing to pay more for them, the market for sustainable innovation is growing in turn. Sustainable innovation assists customers and citizens in managing their lifestyle by enabling them to live happier lives in ways that support sustainable development.

Sustainable innovation provides the foundation for future business; it is not simply reflecting ethical responsibility. The tasks that sustainable innovation is geared towards — the wicked problems — have global significance. For example, the scarcity of clean water is clearly a wicked problem to which answers are being sought everywhere. An all-encompassing way to manage water resources is important, one where the various uses of water are seen as both connected and also as part of the broader environmental context, for example land use patterns.[56] Sustainable innovations have a significant role in the availability and quality of water. To date, they have concerned, for instance, the storage, distribution, pricing, use and reuse of water. The overall, shared goal is to develop sustainable models for water use.

Sustainable innovations can be technical, societal, or systemic. In the case of water, industry invests in technologies that reduce water consumption and reclaim waste-water. Technologies that can help decrease household water consumption and increase waste-water reclamation include water-saving toilets, showers, and faucets. The sustainability of agricultural water consumption is being addressed with the development of methods for "smart" watering, and for better waste-water treatment and reuse. Similarly, the market for other environmental technologies such as solar or wind energy technology is increasing rapidly.[57] In addition, various certification systems and eco-labels have been developed in order to oversee and reduce the pressure put on ecosystems and to incentivise the use of new eco-friendly technologies or to limit the use of scarce resources.

A recent phenomenon related to sustainable innovation is so called *impact investing*.[58] It refers to investments, which aim to reach positive, measurable impacts on social conditions and environment, beside economic return. These investments could be done into firms, funds, bonds or projects. Tools include equities and loans. Two examples, Social impact bonds in UK and DBL Investors in California, US, are presented below.

> "Social impact bonds (SIBs) are designed to help reform public service delivery. SIBs improve the social outcomes of publicly funded services by making funding conditional on achieving results. Investors pay for the project at the start, and then receive payments based on the results achieved by the project. Rather than focusing on inputs (e.g., number of doctors) or outputs (e.g., number of operations), SIBs are based on achieving social 'outcomes' (e.g., improved health). The outcomes are predefined and measurable."
>
> (https://www.gov.uk/social-impact-bonds, referred 17.6.2014)

> "DBL Investors uses venture capital to accelerate innovation in a way that positively affects an organization's social impact as well as its financial success. In fact, we believe the two — positive social change and a healthy financial performance — are inherently connected. DBL invests in and helps nurture outstanding entrepreneurs and companies in Cleantech, Information Technology, Health Care, and Sustainable Products and Services."
>
> (http://www.dblinvestors.com/about/, referred 17.6.2014)

## 2.3. *Systemic change and innovation*

Generally, innovation or its impact is difficult to predict, although favourable conditions can be created to encourage them to emerge. Leaders at national, regional, and organisational levels often are challenged by this reality because establishing these conditions typically require long-term, widespread systemic changes.[59] Similarly, solving wicked problems in a sustainable way requires a systemic view.

We face systemic change and systemic innovations in many challenges and wicked problems of modern society: energy issues, transportation systems, healthcare systems, reforms in agriculture, waste systems etc. Systemic innovations are related to changes in socio-technical systems and often they are described as leaps or transitions. Transitions can be large such as transition from rural to industrial society or more specific such as transition from telegraph to telephone. It is important that systemic innovations are related not only to technological change but also to societal and cultural changes: changes in user contexts and symbolic meanings. In addition, systemic innovation often forms the core of a national innovation strategy. However, these strategies lack practical measures and guidelines. This might result from the concept of systemic innovation being quite new and unclear.[60,61]

An important case of systemic innovation is that of social innovation. Social innovations are defined to refer to "changes in the cultural, normative or regulative structures of the society which enhance its collective power resources and improve its economic and social performance".[62] According to this systemic approach, the creation and development of educational systems, healthcare systems, transportation systems etc. are typical social innovations. Quite often this kind of structural social innovations are created by top–down approach and they presuppose governmental intervention such as legislation and regulation. However, there are also social innovations, which are more directly related to activities of citizen such as new forms of local initiatives, mutual aid, peer support and co-ops. This kind of social innovation stresses the active participation of ordinary citizens and might be called "soft innovation".[63] Soft innovations are often developed by bottom–up approach and in people–public–private partnership. Governments are not the only supporting instance of social innovations, but social enterprises have taken active role to collaborate with citizens to solve local and wicked problems.[64]

A few approaches help in understanding systemic innovation. Firstly, the complexity of social changes such as the development of civil society or the slowing of population growth in developed countries must be understood. Equally important is exploring the adoption of new technologies such as new energy technologies, electric cars, high-speed rails, or electronic databases in the healthcare system. In the case of new technology, this would involve evaluating the technology's maturity, costs, resulting changes in legislation, and so forth. Finally, it is important to understand that the acceptance of the system is affected by the general values of society and the development of national and international trends such as the awareness of climate change and sustainable development. As a whole, systemic innovation includes changes in the market, in the consumer behaviour, in politics, and in the culture.[65,66]

In economics, the theory of complementarities provides a way to analyse systemic innovation.[67,68] According to the theory, two practices are complementary when the advantage of one is greater if both the practices are present. For example, implementing a new information system is more beneficial if relevant training is available. Milgrom and Roberts have examined complementary actions in systems and found that an organisation benefits more from adopting complementary practices than solitary "best practices".[69] The systemic view is useful also in the case of an individual innovation. For example, the widespread adoption of electric cars requires innovations both in technical details (battery capacity, etc.) and in creating a network of charging stations. Parallel, complementary innovations secure the successful implementation of innovation only when working simultaneously.

The dynamics of change provides interesting insights into sustainable innovation. The increasing complexity of society and wicked problems highlight the need for systemic innovation. There is a need for action models that include services, operational processes, organisational structures, value chains and technological changes. As noted in the previous section, the changes essential to innovation that are actualised simultaneously in different sectors and arenas, as well as the impacts of innovation, need to be assessed carefully. Mitsuru Kodama, who studied the structural reforms of Matsushita Electric, a traditional Japanese manufacturer, stressed how the most effective development requires not only optimising the individual parts but also redesigning the whole system so that the actors can enhance the synergy of overall optimisation.[70] Synergy facilitates solving wicked problems.[71]

In many ways, systemic innovation is a systemic learning process, as are the processes of structural change and wicked problem solving. Through innovation, the system is adjusted and its operations and structures changed. Likewise, innovation is developed through feedback and experience. Sustainable innovation is systemic by nature, thus it takes into account a number of complementary technological, organisational, and operational changes. A single innovator seldom has all the know-how required to successfully implement innovations. Actors in the innovation ecosystem depend on the resources, expertise, and connections of other actors. Innovation, therefore, requires close cooperation and interaction.[72,73] Especially when the emphasis is on the impact of innovation and on the expected 'good' that innovation creates, the systemic approach is needed. As shown earlier, the actual impact depends on the system and on how different parts are linked with each other.

## 2.4. *Inclusive innovation*

One defining attribute of sustainable innovation is *inclusivity*, reflecting the fact that innovations emerge from a synthesis when different knowledge is combined. In the first decade of the 21st century, innovation researchers have emphasised networked, open, and diverse forms of innovation.[74] Inclusive innovation implies that all people should have the opportunity to use their potential to seek creative solutions to the challenges they deem important. A background for this argument is the notion that the most important resources of innovation are creative, skilled people, both in the workplace and in everyday life. Inclusive innovation could be summarised by the principle "innovation for all". This means not only that all people must have some opportunities to innovate but also that innovation must serve and benefit all people.

Inclusive innovation supports collective wisdom and the crowdsourcing of problems.[75,76] This kind of development and other forms of mass collaboration have a deep impact on the economy, business and the government.[77] In a deeper sense, sustainable and inclusive innovation promotes new forms of democracy, where citizens have the right and the opportunity to be creative and to contribute to improvements in services, products and the structure of public organisations like municipalities, schools and hospitals.[78]

Generally inclusive innovation has its background in one of the most important changes in today's innovation: The shift from closed innovation models towards open and distributed models.[79] The open innovation paradigm assumes that firms can and should use external ideas as well as internal ideas, and internal and external paths to market. In open innovation, ideas and new technology outside the company are constantly looked for. New ideas can be introduced at any stage of the product development. Well-known examples of open innovation utilisers are IBM, Procter & Gamble, and Intel in the United States; the Philips Group Corporation and Nokia in Europe; and Omron and Sunimoto 3M in Japan.

von Hippel emphasised the importance of users, who include more than just customers.[80] Companies, organisations, cities and municipalities are also users. Users often want to develop products and services to better meet their individual needs. Almost every organisation wants to tailor made software according to their needs, by themselves or by ordering additional services. Companies modify the machinery and equipment they purchase, improve their properties, connect them to other devices and so on. Individual consumers do the same with the household equipment or software they buy.

Open innovation is taking place in the market place: It's selling and purchasing ideas. On the other hand, public innovation activities are usually organised outside companies and for free. For example open-source products like Linux system have been created by voluntary actions outside the formal markets. Open-source products are commons or public goods as opposed to products owned by private companies, and the property in open source is configured fundamentally around the right to distribute, not the right to exclude.[81] Chesbrough argues also for open "business models".[82] For example, Red Hat has developed tools that facilitate the installation and operation of Linux on different computers, and makes profit by selling these software and services.

In between open and public innovation are different innovation alliances. Alliances are group of companies cooperating in order to produce innovations.[83] The idea is in sharing and combining information without limiting the cooperation to any particular technology or case. What is common to these alternatives modes is that the innovation happens in environment where the needed skills and resources are scattered in different places.

To sum up, inclusive innovation is a multidimensional approach to product and service development and implementation. On the development side, it facilitates opening innovation processes to global cooperation; possessing the ability to find the best ideas and experts from all around the world; building up and controlling various innovation networks; developing the ability to transform all investments, externalisations, and acquisitions into learning and innovation processes. On the implementation side, inclusive innovation facilitates learning the tacit knowledge of various markets and cultures, and properly managing diversity within the company and with partner relations and markets.[84]

## 3. Conclusions: Innovation Policy for Sustainable Innovation

Throughout this chapter, we have characterised sustainable innovation as inclusive, as systematic, and as motivated by the need to solve wicked problems. We have emphasised the impact of innovation and innovative solutions for wicked problems.

In the end, we argue for innovation policies — at the company and the national levels — that shift towards sustainable, inclusive, and systemic innovation. The aim of the chapter has been to awaken thinking about innovation activity and its development in new ways. Traditionally, the goals of innovation policies all over the world have been economic growth and the growth of productivity. Policies have been technologically emphasised, innovation processes closed and institutions isolated. A national innovation system (NIS) has been the basic concept of innovation policy and the national level has dominated the discourse.

Nowadays innovation strategies in many countries are adopting broad-based policies and call for cooperative and user centred approach of innovation.[85–87] However, a more radical leap is needed in order to view the innovation policy from a new perspective.

Sustainable innovation policy takes sustainable well-being and sustainable development as the basic values, leaving economic growth with an instrumental value. It also shifts the dominance in the discussion and focus from a national level to both local and global levels when the basic field of innovation activities will be different innovation ecosystems and not the national innovation system. It takes culture and creativity as essential part of innovation environment and aims for spontaneous processes and radical innovations. As there is a need to move from national level to places where people work together, in other words, to local ecosystems, where sustainable innovation policy is localised and where people and their networks serve as the primary sources for innovation activities.[88] Understanding people and the flow of ideas as a basis of innovation activities challenges the traditional innovation policy, and requires systemic approach and deep institutional cooperation and interaction.[89]

The characterisation above is simplified but it provides a framework from which one can consider the current state and rhetoric on innovation policies. The next step after the characterisation of sustainable innovation is focusing on the capacities needed to support sustainability in innovation. The proposed emphasis on human, environmental and societal aspects in addition to economical success is quite new and requires more studies. Despite the new open paradigms of innovation, innovation as a whole is typically considered as a top-level activity that does not include citizens. This chapter argues that producing human-centred solutions for wicked problems is not possible if majority of people are out of reach of innovation activities. In the sustainable innovation policy, all innovation activities are considered in terms of how they contribute to good life and solving wicked problems such as climate change, poverty, aging society, polarisation or illiteracy. Sustainable innovation outlines significant changes in mindsets: All the effects of innovation must be evaluated according to their contribution to good life.

## References

1. Harris, J. M., Wise, T. A., Gallagher, K. P. and Goodwin, N. R. (2001). *A Survey of Sustainable Development* (Island Press, Washington, London).
2. Rittel, H. W. J. and Webber, M. M. (1984). Planning problems are wicked problems. In Cross, N. (ed.), *Developments in Design Methodology* (John Wiley and Sons, Chichester), pp. 135–144.
3. Kao, J. (2007). *Innovation Nation, How America is Losing Its Innovation Edge, Why That Matters, and What We Can Do to Get It Back* (Free Press, New York, London, Toronto, Sydney).
4. Carlson, C. R. and Wilmot, W. W. (2006). *Innovation: The Five Disciplines for Creating What Customers Want* (Crown Business, New York).
5. Auerswald, P. (2012). *The Coming Prosperity: How Entrepreneurs Are Transforming the Global Economy* (Oxford University Press, Oxford).
6. Lockwood, T. (ed.). (2010). *Design Thinking. Integrating Innovation, Customer Experience, and Brand Value* (Allworth, New York).
7. van Bueren, E., Klijn, E. and Koppenjan, J. F. M. (2003). Dealing with wicked problems in networks: Analyzing an environmental debate from a network perspective. *Journal of Public Administration Research and Theory*, 13: 193–212.
8. Weber, E. P. and Khademian, A. M. (2008). Wicked problems, knowledge challenges, and collaborative capacity builders in network settings. *Public Administration Review*, 68(2): 334–349.
9. Wexler, M. N. (2009). Exploring the moral dimension of wicked problems. *International Journal of Sociology and Social Policy*, 29(9/10): 531–542.
10. Luoma-Aho, V. and Vos, M. (2010). Towards a more dynamic stakeholder model: Acknowledging multiple issue arenas. *Corporate Communications: An International Journal*, 15(3): 315–331.
11. Schumpeter, J. (1934). *The Theory of Economic Development: An Inquiry into Profit, Capital, Credit, and Business Cycles* (Harvard University Press, Cambridge, Mass.).
12. Carlson, C. R. and Wilmot, W. W. in Ref. [4].
13. Crespi, G. and Zuniga, P. (2012). Innovation and productivity: Evidence from six latin American countries. *World Development*, 40(2): 273–290.

14. Hall, R. and Jones, C. (1999). Why do some countries produce so much more output per worker than others? *The Quarterly Journal of Economics*, 114(1): 83–116.
15. Mumford, M. D. (2002). Social Innovation: Ten cases from Benjamin Franklin. *Creativity Research Journal*, 14(2): 253–266.
16. Jensen, M. B., Johnson, B., Lorenz, E. and Lundvall, B.-Å. (2007). Forms of knowledge and modes of innovation. *Research Policy*, 36(5): 680–693.
17. Chesbrough, H. (2003). *Open Innovation, The new Imperative for Creating and Profiting from Technology* (Harvard Business School Press, Boston, Mass.).
18. Chesbrough, H. (2006). *Open Business Models, How to Thrive in the New Innovation Landscape* (Harvard Business School Press, Boston, Mass.).
19. von Hippel, E. (2005). *Democratizing Innovation* (The MIT Press, Cambridge).
20. Chesbrough, H. (2011). *Open Services Innovation, Rethinking Your Business to Grow and Compete in a New Era* (Jossey-Bass, San Francisco).
21. Benkler, Y. (2006). *The Wealth of Networks. How Social Production Transforms Markets and Freedom* (Yale University Press, New Haven).
22. Mulgan, G. (2007). *Ready or Not? Taking Innovation in the Public Sector Seriously*. NESTA Provocation 03 (NESTA, London).
23. Altenburg, T., Stamm, A. and Schmitz, H. (2008). Breakthrough? China's and India's transition from production to innovation. *World Development*, 36(2): 325–344.
24. Radjou, N., Prabhu, J. and Ahuja, S. (2012). *Jugaad Innovation: Think Frugal, Be Flexible, Generate Breakthrough Growth* (Jossey-Bass, San Francisco).
25. Porter, J., Joseph, M., Seppä, M. and Biswas, D. (2010). Reflection of the GVL Live Case Approach: Integrating Action Learning, Research and Business for Co-Creating Global Enterprises. A Finnish Live Case Study on the Indian Health Kiosk Venture. *EBRF Conference Proceedings*, September 2010 (Nokia, Finland).
26. Drejer, I. (2004). Identifying innovation in surveys of services: A Schumpeterian perspective. *Research Policy*, 33(3): 551–562.
27. Prahalad, C. K. and Ramaswamy, V. (2003). New frontier of experience innovation. *MIT Sloan Management Review*, 44(4): 12–18.
28. Conklin, J. (2006). *Wicked Problems and Social Complexity. Dialogue Mapping: Building Shared Understanding of Wicked Problems* (Wiley, Hoboken).
29. Denning, P. J. and Yaholkovsky, P. (2008). Getting to "we". *Communications of the ACM*, 51(4): 19–24.
30. Brandbury, J. A. A. (1989). *Product Innovation — Idea to Exploitation* (John Wiley and Sons, Chichester).
31. Koen, P., Ajamian, G., Burkart, R., Clamen, A., Davidson, J., Amore, R. D., Elkins, C., Herald, K., Incorvia, M., Johnson, A., Karol, R., Seibert, R., Slavejkov, A. and Wagner, K. (2001). Providing clarity and a common language to the "Fuzzy front End". *Research–Technology Management*, 44(2): 46–55.

32. Padmore, T., Schuetzea, H. and Gibson, H. (1998). Modeling systems of innovation: An enterprise-centered view. *Research Policy*, 26(6): 605–624.
33. Schoen, J., Mason, T. W., Kline, W. A. and Bunch, R. M. (2005). The innovation cycle: A new model and case study for the invention to innovation process. *Engineering Management Journal*, 17(3): 3–10.
34. Rogers, E. M. (2003). *Diffusion of Innovations*, 5th Ed. (The Free Press, New York).
35. Crossan, M. M. and Apaydin, M. (2010). A multi-dimensional framework of organizational innovation: A systematic review of the literature. *Journal of Management Studies*, 4(6): 1154–1191.
36. Narayanan, V. K. (2001). *Managing Technology and Innovation for Competitive Advantage* (United States of America, Prentice-Hall Inc.).
37. Oksanen, K. and Hautamäki, A. (2014). Transforming regions into innovation ecosystems — A model for renewing local industrial structures. *The Innovation Journal: The Public Sector Innovation Journal*, 19(2), Article 5.
38. Medina, C. C., Lavado, A. C. and Cabrera, R. V. (2005). Characteristics of innovative companies: A case study of companies in different sectors. *Creativity and Innovation Management*, 14: 272–287.
39. Laird, D. M. (2005). Organizational culture's influence on creativity and innovation: A review of the literature and implications for human resource development. *Advances in Developing Human Resources*, May 1, 7: 226–246.
40. Puro, J.-P. (2002). Finland: A mobile culture. In Katz, J. E. and Aakhus, M. A. (eds.), *Perpetual Contact: Mobile Communication, Private Talk, Public Performance* (Cambridge University Press, Cambridge), pp. 19–29.
41. Katz, J. E. and Aakhus, M. A. (eds.). (2002). *Perpetual Contact: Mobile Communication, Private Talk, Public Performance* (Cambridge University Press, Cambridge).
42. Sethi, S. P. (1994). *Multinational Corporations and the Impact of Public Advocacy on Corporate Strategy: Nestle and the Infant Formula Controversy. Issues in Business Ethics* (Kluwer Academic, Boston).
43. Rogers, E. M. in Ref. [34].
44. Rogers, E. M. and Shoemaker, F. F. (1971). *Communication of Innovations. A Cross-Cultural Approach* 2nd Ed. (The Free Press, New York).
45. Simula, H. (2011). *Management of Commercialization. Case Studies of Industrial, Business-to-Business Product Innovations.* Aalto University School of Science, Department of Industrial Engineering and Management, Doctoral Dissertation Series (Aalto University, Espoo).
46. Smith, D. (2006). *Exploring Innovation* (McGraw Hill, UK).
47. Helpman, E. (2004). *The Mystery of Economic Growth* (Harvard University Press, Cambridge, Mass.).
48. Nidumolu, R., Parahalad, C. K. and Rangaswami, M. R. (2009). Why sustainability is now the key driver of innovation. *Harvard Business Review*, September.

http://hbr.org/2009/09/why-sustainability-is-now-the-key-driver-of-innovation/ar/5, referred 29.10.2012.

49. Hautamäki, A. (2010). *Sustainable Innovation. A New Age of Innovation and Finland's Innovation Policy*. Sitra Reports 87 (Sitra, Helsinki).
50. Hall, J. and Clark, W. (2003). Introduction to the special issue on environmental innovation. *Journal of Cleaner Production*, 11: 343–346.
51. Boons, F. and Lüdeke-Freund, F. (2013). Business models for sustainable innovation: State of the art and steps towards a research agenda. *Journal of Cleaner Production*, 45: 9–19.
52. Bhatti, Y. A. and Ventresca, M. (2012). *The Emerging Market for Frugal Innovation: Fad, Fashion, or Fit?* January 15. Available at SSRN http://dx.doi.org/10.2139/ssrn.2005983, referred 20.12.2012.
53. Radiou, N., *et al.* in Ref. [24].
54. Porter, M. and Kramer, M. (2011). Creating shared value. How to reinvent capitalism- and unleash the wave of innovation and growth. *Harvard Business Review*, 89(1/2): 62–77.
55. Porter, M. (1980). *Competitive Strategy* (Free Press, New York).
56. Bergkamp, G. and Sadoff, C. (2008). Water in a Sustainable Economy. In *State of the World 2008, Innovations for a Sustainable Econom Sustainable Economy* (Worldwatch Institute, Washington), pp. 107–238.
57. Verdantix. (2012). *Global Energy Leaders Survey 2012: Emerging Economies* (Verdantix Ltd., London).
58. Bannick, M. and Goldman, P. (2012). *Priming the Pump: The Case for a Sector Based Approach to Impact Investing.* http://www.omidyar.com/pdf/Priming_the_Pump_Sept_2012.pdf.
59. Geels, F. W. and Schot, J. W. (2007). Typology of sociotechnical transition pathways. *Research Policy*, 36(3): 399–417.
60. Hämäläinen, T. and Heiskala, R. (eds.). (2007). *Social Innovations, Institutional Change and Economic Performance. Making Sense of Structural Adjustment Processes in Industrial Sectors, Regions and Societies* (Edward Elgar, Cheltenham).
61. Streeck, W. and Thelen, K. (eds.). (2005). *Beyond Continuity: Institutional Change in Advanced Political Economies* (Oxford University Press, Oxford).
62. Heiskala, R. (2007). *Social innovations: Structural and power perspectives.* In Hämäläinen, T. and Heiskala, R. (eds.) in Ref. [61], pp. 52–79.
63. Oksanen, K. and Hautamäki, A. (2014) in Ref. [37].
64. Auerswald, P. (2012). *The Coming Prosperity: How Entrepreneurs Are Transforming the Global Economy* (Oxford University Press, Oxford).
65. Geels, F. W. (2010). Ontologies, socio-technical transitions (to sustainability), and the multi-level perspective. *Research Policy*, 39(4): 495–510.
66. Geels, F. W. and Schot, J. W. (eds.) in Ref. [60].

67. Brynjolfsson, E. and Saunders, A. (2010). *Wired for Innovation — How Information Technology Is Reshaping the Economy* (The MIT Press, Cambridge).
68. Teece, D. J. (1987). Profiting from Technological Innovation: Implications for Integration, Collaboration, Licensing, and Public Policy. In Teece, D. J. (ed.), *The Competitive Challenge* (Ballinger Pub. Co., Cambridge Mass.), pp. 185–219.
69. Milgrom, P. and Roberts, J. (1995). Complementarities of fit: Strategy, structure, and organizational change. *Journal of Accounting and Economics*, (19): 179–208.
70. Kodama, M. (2007). Innovation through boundary management — A case study in reforms at Matsushita electric. *Technovation*, 27(1–2): 15–29.
71. van Bueren, E., Klijn, E. and Koppenjan, J. F. M. (2003). Dealing with wicked problems in networks: Analyzing an environmental debate from a network perspective. *Journal of Public Administration Research and Theory*, 13: 193–212.
72. Kosonen, M. and Doz, Y. (2008). *Fast Strategy: How Strategic Agility Will Help You Stay Ahead of the Game* (Pearson Education Ltd., Harlow).
73. Scott, A. and Storper, M. (2003). Regions, globalization, development. *Regional Studies, Taylor and Francis Journals*, 37(6–7): 579–593.
74. Chesbrough, H. in Ref. [17], Chesbrough, H. in Ref. [18], Chesbrough, H. in Ref. [20], E. von Hippel in Ref. [19].
75. Surowiecki, J. (2004). *The Wisdom of Crowds: Why the Many Are Smarter Than the Few and How Collective Wisdom Shapes Business, Economies, Societies and Nations* (Doubleday, Garden City).
76. Weinberger, D. (2011). *Too Big to Know* (Basic Books, New York).
77. Tapscott, D. and Williams, A. D. (2007). *Wikinomics: How Mass Collaboration Changes Everything* (Penguin, New York).
78. Benkler, Y. in Ref. [21], von Hippel, E. in Ref. [19].
79. Chesbrough, H. in Ref. [17], Hautamäki, A. in Ref. [49].
80. von Hippel, E. in Ref. 19.
81. Weber, S. (2004). *The Success of Open Source* (Harvard University Press, Boston, Mass.).
82. Chesbrough, H. in Ref. [18].
83. Miles, R. E., Miles, G. and Snow, C. C. (2005). *Collaborative Entrepreneurship: How Communities of Networked Firms Use Continuous Innovation to Create Economic Value* (Stanford University Press, Stanford).
84. Nonaka, I. and Takeuchi, H. (1995). *The Knowledge–Creating Company: How Japanise Companies Create the Dynamics of Innovation* (Oxford University Press, New York).
85. Etzkowitz, H. and Leyersdorf, L. (2000). The dynamics of innovation: From National systems and "Mode 2" to a triple helix of university–industry–government relations. *Research Policy*, 29(2): 109–123.

86. Lundvall, B.-Å. Johnson, B. Andersen, E. S. and Dalum, B. (2002). National systems of production, innovation and competence building. *Research Policy*, 31(2): 213–231.
87. Hautamäki, A. in Ref. [49].
88. *Ibid*.
89. Seshadri, S. and Shapira, Z. (2003). The flow of ideas and timing of evaluation as determinants of knowledge creation. *Industrial and Corporate Change*, 12(5): 1099–1124.

# Chapter 6

# FUTURES OF INNOVATION SYSTEMS AND INNOVATION MANAGEMENT: OPEN INNOVATION PARADIGM ANALYSED FROM FUTURES PERSPECTIVES

Jari Kaivo-oja

*Finland Futures Research Centre*
*Turku School of Economics,*
*University of Turku, Yliopistonkatu 58 D*
*33100 Tamper*
*Finland*

Teemu Santonen

*Laurea University of Applied Sciences*
*Vanha maantie 9, 02650 Espoo, Finland*
*teemu.santonen@laurea.fi*

This chapter provides a futures-oriented analysis of Open Innovation (OI) paradigm The future of OI paradigm depends on the future of CI paradigm and vice versa. That is logical reason why they introduce the concept of symbiotic innovation to innovation management literature.

In this chapter, authors elaborate open innovations as efficiency drivers in evolutionary economic processes (Section 2). In Section 3, authors introduce the open innovation paradigm and key concepts of futures-oriented open innovation management. In Section 4, authors present a scientific discussion about systemic change and the critical role of individuals in innovation management. Authors pay special attention on the difference between open and closed innovation process from an individual's point of view. In Section 5, authors introduce the Quartet Helix Model (as additional model to Triple Helix Model) and an analysis of new dynamics of open innovation processes. In Section 6, authors elaborate and discuss open innovation management roadmaps to the future. Section 7 provides elaborations of alternative innovation models for private enterprises, with special attention to foresight–innovation linkages. In Section 8, authors analyse links between radical innovation management and OP paradigm. Section 9 provides summary of key insights and results.

*Keywords*: Futures of Innovation Systems; Quartet Helix Model, Open Innovation Paradigm; Open Innovation; Closed Innovation; Incremental Innovation; Radical Innovation; Symbiotic Innovation.

## 1. Introduction

Over the last decades, innovations have generally been discussed within the terms of closed innovation management, in which firms have conducted their innovation efforts via proprietary models and developed products and services through internal research and development (R&D) activities. However, for the past 10 years, it has been argued that the economic system is entering a new era of Open Innovation (OI)[1], which has received a lot of scientific attention from the empirical and theoretical economic literature. According to the conventional definition of OI, it is "the use of purposive inflows and outflows of knowledge to accelerate internal innovation, and expand the markets for external use of innovation, respectively".[2]

There are many driving forces and trends behind the OI paradigm (see p. 64).[3] These include fast technological change and increasing complexity of products[4], high innovation speed[5], shortened product lifecycles[5], spread of knowledge in the value chain and concentration on core competences[6–7] and business models that integrate across various industries.[8–9] To our knowledge, there are very few weak signals which would indicate serious anti-trends for these trends. Therefore, we can assume that also for the forthcoming years the futures of the OI approach are strongly dependent on the futures of business environment driven by the above-presented trends. How forceful these trends will be in the future defines also the futures of OI management. Typically, as firms' internal resources are not sufficient to meet these new challenges, they have to access external resources. The driving force of OI is based on the fact that many firms are obliged to innovate and develop new products under extremely tough resource constraints and timelines in order to stay in the market and keep themselves competitive. This is also the logical theoretic reason for small companies, when compared to large corporations, being often more eager to adopt OI.[10–12] This would mean that large corporations are not interested in OI management.

The open innovation model can be seen as a call to return to the late 19th and early 20th century model of innovation with its diverse and rich market for technology and small, externally-oriented R&D labs and networked teams.[13,14] The old vertically integrated industrial R&D model promulgated by Chris Freeman and other remarkable scholars[15–18] is still relevant, but open innovation is increasingly seen as the most relevant approach in the literature of OI. Thus, we agree that the CI and OI approaches are not substitutes but complementary, and we expect this be the true analysis in the future, too.

The aim of this chapter is to elaborate the futures of innovation systems, with a special attention to the open innovation paradigm and its possible development paths in scientific discussion. We think that, in many ways, the open innovation approach is rooted to its current scientific pre-conditions and research activities and, therefore, the futures of the open innovation paradigm are linked to the current trends of research in this field. The futures of the open innovation paradigm, of course, are also dependent on the closed innovation paradigm and its popularity among companies and corporations. Therefore, we must note also some of the key findings in the CI paradigm in order to predict the future of open innovation. As pointed out in this chapter, our special focus is on the linkage between innovation research and futures orientation of innovation management activities. Our chapter thus includes some theoretical diagnoses of innovation management, some future-oriented prognoses of the open innovation paradigm and, finally, some policy-relevant conclusions for global innovation management and leadership.

This chapter is organised as follows: In Section 2, we elaborate open innovations as efficiency drivers in evolutionary economic processes. In Section 3, we introduce the open innovation paradigm and key concepts of futures-oriented open innovation management. Section 4 presents a scientific discussion about systemic change and the critical role of individuals. We pay special attention on the difference between open and closed innovation process from an individual's point of view. In Section 5, we introduce the Quartet Helix model and new dynamics of open innovation processes. In Section 6, we elaborate and discuss open innovation management roadmaps to the future. Section 7 elaborates alternative innovation models for private enterprises, with special attention to foresight–innovation linkages. In Section 8, links between radical innovation management and OP paradigm is discussed. In Section 9, we draw conclusions from the key theoretical OI insights and OI analyses.

## 2. Open Innovations as Efficiency Drivers in Evolutionary Economic Processes

OI has a special role in networked market economies. OI theory can be connected theoretically to evolutionary theory of innovation.[19] A paradigm-based theory of production[20–22] predicts that, in general, in the space of unit inputs, micro-coefficients are distributed somewhat in the way depicted in Figure 1. In the figure, X1 and X2 are inputs (typically, capital X1 and labour X2). Thus technological progress usually means that firms need less input to produce more output.

Following Dosi and Nelson[23] (p. 83), suppose that at time $t$ the coefficients are $C_1$, ..., $C_n$, where 1, ..., $n$ are the various techniques

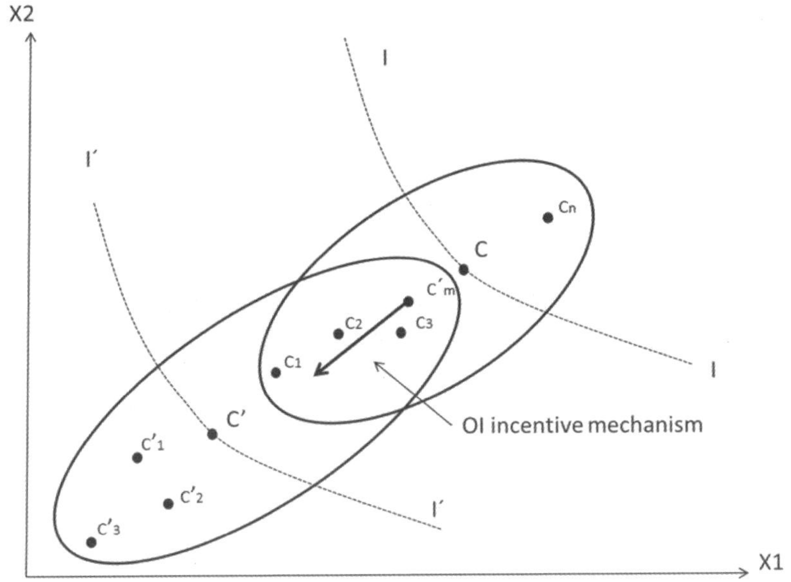


**Figure 1.** OI incentive mechanism. Micro-heterogeneity and technological trajectories (modified, see [Ref. 23, p. 83])

labelled in order of decreasing efficiency at time *t*. It is straightforward, for example, that technique $C_1$ is unequivocally superior to the other ones, no matter what the relative prices are: It can produce the same unit output, with less input volume of both $X_1$ and $X_2$. Inputs $X_1$ can be, for example capital and input $X_2$ labour. The same applies to the comparison between technology option $C_3$ and $C_n$ (see more,[23] pp. 82–85).

Let us assume that technology $c_1$ has been managed through OI management. This aspect of $c_1$ makes this particular technology option a very attractive technology option for firms. Economic incentive to adopt this superior OI technology will be very high compared to other CI technology options, not only because it allows for a firm to have the same unit output with less inputs $X_1$ and $X_2$. This potential advantage of OI management has not received much theoretical attention in the microeconomic literature, but, if this accelerating incentive mechanism works as presented in Figure 20, OI management can lead us to much faster productivity gains than it did before it was introduced as a tool of innovation management.

Once this theoretical and comparative advantage aspect of OI management is understood broadly among stakeholders, it may in future motivate stakeholders to adopt the OI paradigm in many organisations — just because it leads to better productivity in regional and national economies. Of course, the microeconomics of OI management needs more theoretical attention in the future.

## 3. Open Innovation Paradigm and Key Concepts of Futures-Oriented Open Innovation Management

In Table 1, we have outlined the generations of innovation management and futures research (see especially p. 64 in Ref. [31]).[3,24,25] The development of OI management has been developed as a part of Generation 3 and 4. The development of CI has been developed as a part of Generation 1 and 2. Historically, the borderlines of generations of innovation processes are not so clear-cut (see e.g., Ref. [26]). We can claim that the development of OI can be linked more closely to coupled innovation processes and innovation in systems or networks.[27,28] Networked foresight and exploratory futures research are also linked to the development of OI management. In a way, OI management is quite strongly linked to the development network economy. OI futures are probably very closely linked to the changing behaviour of networks, crowds and markets. The level of turbulence in these systems will increase OI activity. Thus, the future of OI is linked to networks, markets and crowds, but mostly on networks and crowds.[28,29]

**Table 1.** Generations of innovation management and futures research[33]

| | Innovation processes | Futures research |
|---|---|---|
| Generation 1 | Technology push | Technology forecasting |
| Generation 2 | Market pull | Technology assesment |
| Generation 3 | Coupled innovation processes | Exploratory futures research |
| Generation 4 | Innovation in systems or networks | Networked foresight |

For a generalised innovation paradigm, it is important to define it as a complementary element of the closed innovation paradigm. Together closed and open innovation management thinking create the core approaches of innovation management. Closed innovation approach relies on (1) authority-based hierarchy or (2) consensus-based hierarchy. Open innovation approach relies on (1) markets and contracts, (2) partnerships, alliances and corporate venture capital (CVC), (3) contents, tournaments and platforms and (4) users and communities (Table 2). As we see, there are many differences between communication channels, incentives and management of property rights in these innovation management approaches[30] (pp. 917–919).

In Figure 2, we see the visualisation of the open innovation model. The model includes two important elements: R&D operations and commercialisation. In the model, internal and external innovation activities are combined. External R&D activities are typically: (1) External research projects, (2) venture investments, (3) licensing-in operations and (4) licensing-out operations. External commercialisation activities are: (1) Outsourcing-in operations, (2) all kinds of partnership networking, (3) other firms' market and business models, (4) own firms' new market and business models and (5) outsourcing partnership networking[31] (p. 207). All these OI operations are linked to new markets.

In Figure 3, we have described the 3 basic models of OI activities: (1) Technology exploration, (2) technology exploitation and (3) other IO activities. This figure includes key OI activities.

OI innovations emerge in the conditions of technological and market turbulence. Technological turbulence and market turbulence can have direct and indirect impacts on OI management. A turbulent technological environment makes current technology obsolete and requires new ones for its development. Typically, it is expected that in a volatile technological environment companies tend to acquire greater external technology because such an environment rapidly causes its current technological knowledge and products to be become obsolete.

Market turbulence, which quickly makes company's market knowledge obsolete, refers to the degree of variation in customer preference and customer demand. Firms must take actions to meet

**Table 2.** Open and closed innovation management[30] (p. 918)

| | **Firm/closed innovation** | | **Open innovation** | | | |
|---|---|---|---|---|---|---|
| | Authority-based hierarchy | Consensus-based hierarchy | Markets/ contracts | Partnerships/ alliances/ CVC | Contents/ tournaments/ platforms | Users/ Communities |
| Communication channels | Vertical, socially embedded within firm | Horizontal, socially embedded within firm | Limited selective innovations | Bilateral, socially embedded | Horizontal, broadcast, IT supported | Horizontal, socially embedded outside the firm |
| Incentives | Low-powered | Low-powered | High-powered | Cooperative, high-powered | Moderately-powered | Low-powered |
| Property rights | Possessed by local firm | Possessed by focal firm | Externally owned and exchanged | Negotiated | Varied (dispersed or focal firm) | None |

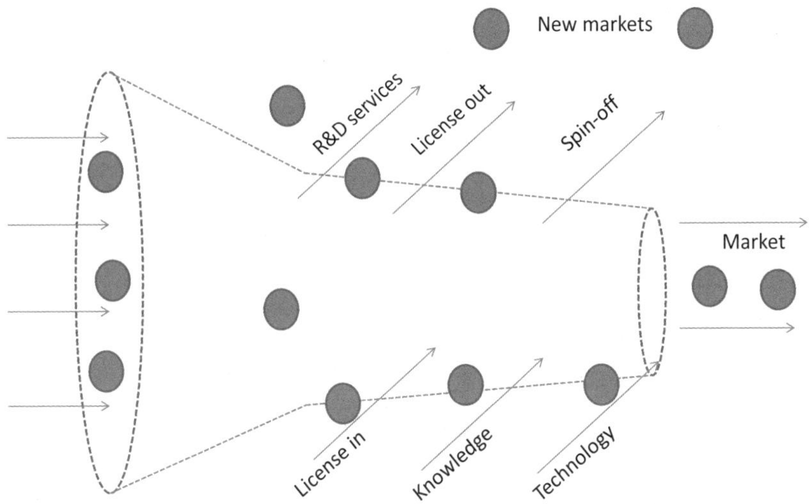


**Figure 2.** Open innovation management model and its key activities[32] (p. 8930)

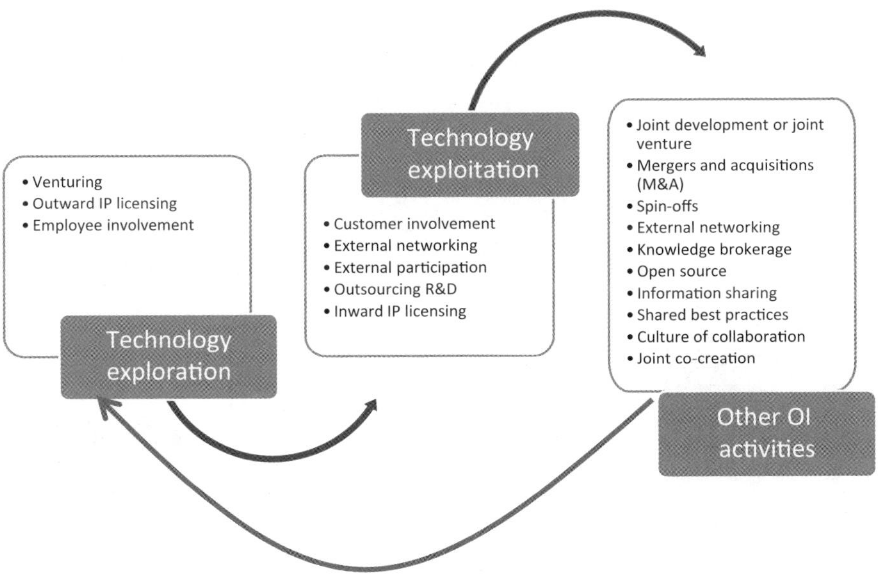


**Figure 3.** Technology exploration, technology exploitation and other OI activities[1–5]

latent customer needs and demands and to sustain competitiveness. In these turbulent conditions, the OI framework often becomes very attractive.

External technology acquisition (ETA, inbound open innovation) and external technology exploitation (ETE, outbound open innovation) are key theoretical concepts in open innovation management. In some cases, turbulence factors strengthen ETA and ETE activities, but they can also weaken those activities. This empirical question needs more empirical research. In Figure 3, we have visualised OI activities[33] (p. 369) which can have economic impacts on firm performance.

We can summarise this figure as follows:

(1) ETA = f(IRD1, TT1, MT1),
(2) ETE = f(ETA1, IR2, TT2, MT2), and
(3) Firm performance = f(ETA, ETE).

Figure 4 explains why so many companies have modified their innovation process towards the Open Innovation paradigm.

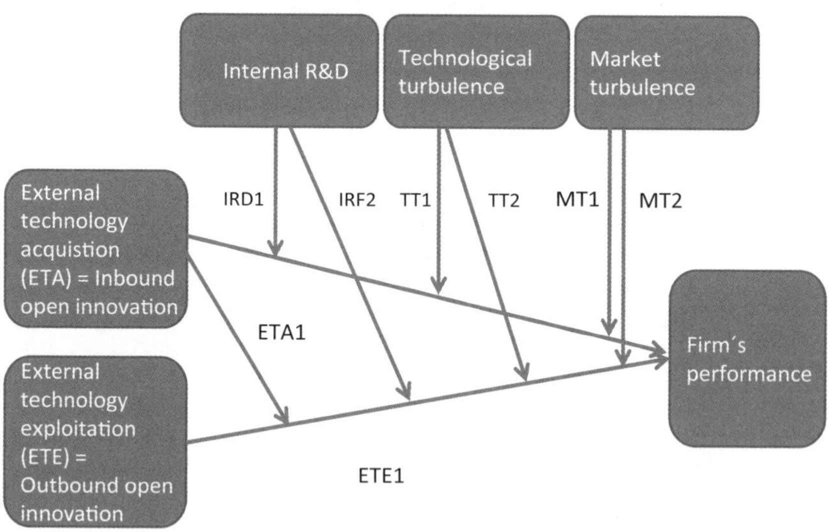


Figure 4. Firm's performance and inbound and outbound open innovation activities

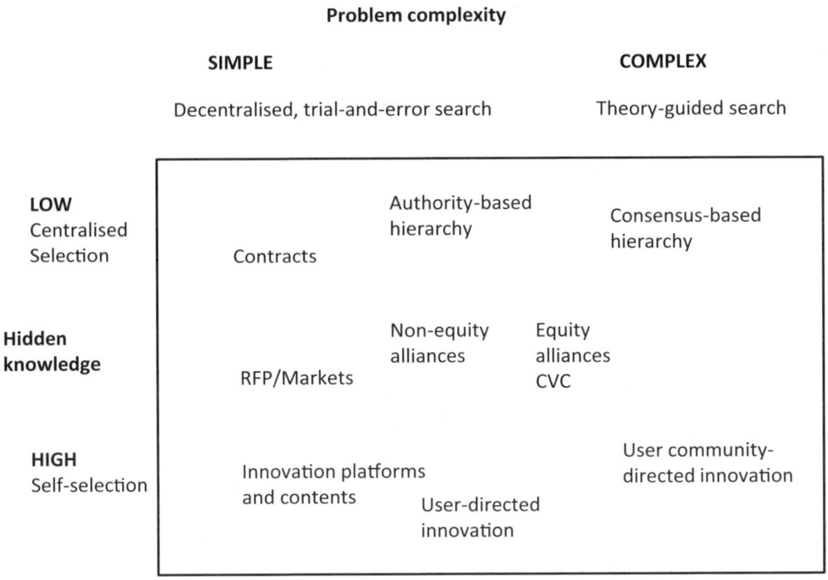


**Figure 5.** Open innovation thinking, problem complexity and hidden knowledge[33] (p. 918)

This model shows how important open innovation activities are for a firm's performance (Figure 4). The perspective of open innovation is linked to a profit-maximising strategy that targets both value creation and value appropriation.[34] The performance of firms depends on OI activities. We can also note that the futures of the open innovation paradigm are closely linked on firms' capacities and capabilities to utilise OI management tools, and vice versa.

One key issue in OI management is the management of complexity and knowledge. Figure 5 provides a visualisation of these two key issues of OI management.

The governance of innovation is always based on knowledge and on complexity of problems. Also open innovation management is based on knowledge. In different problem situations, the optimal solution methods are the ones in Table 3[35] (p. 916).

We can see that OI is typically linked to medium or high hidden knowledge and self-selection processes but also to complex and theory guided-search. CI is more linked to centralised selection and

Table 3. Optimal search methods by alternative problem types[35] (p. 916)

| | | Problem complexity | |
|---|---|---|---|
| | | High | Low |
| Hidden knowledge | High | Open, theory-guided search | Open, directional, trial-and-error search |
| | Low | Centralised, theory-guided search | Centralised, directional, trial-and-error search |

to simple problem complexity. Complexity and high level of self-selection/hidden knowledge are critical factors behind the development of OI.

## 4. Systemic Change and Individuals: Open Vs. Closed Innovation from an Individual's Point of View

From system theoretical perspective, we can present a five-stage model of systemic change. The stages are (1) imitation, (2) inertia, (3) sub-optimisation, (4) change of game, and (5) shift of paradigm. In Figure 6, we have visualised these levels of systemic changes.

Theoretically, in this systemic context, innovative changes start at individual level, and they reach a broader scope via organisations and finally via national and international networks. Without open innovation management, no remarkable changes could occur outside organisations and networks. Figure 6 visualises also the relevance of open innovation itself (see p. 4).[36]

In a systemic change process, imitation is the most basic force. Competitive advantage is, by definition, a distinction. Successful distinctions are eroded by imitation. Change processes start by the elimination of successful distinctions. Inertia is the responsible force for all the second-level tendencies and behaviours that delay human reactions to technological breakthroughs. Inertia creates time lags in response to new breakthroughs. Sub-optimisation, if excessive, destroys a proven competitive advantage. In sub-optimisation, the

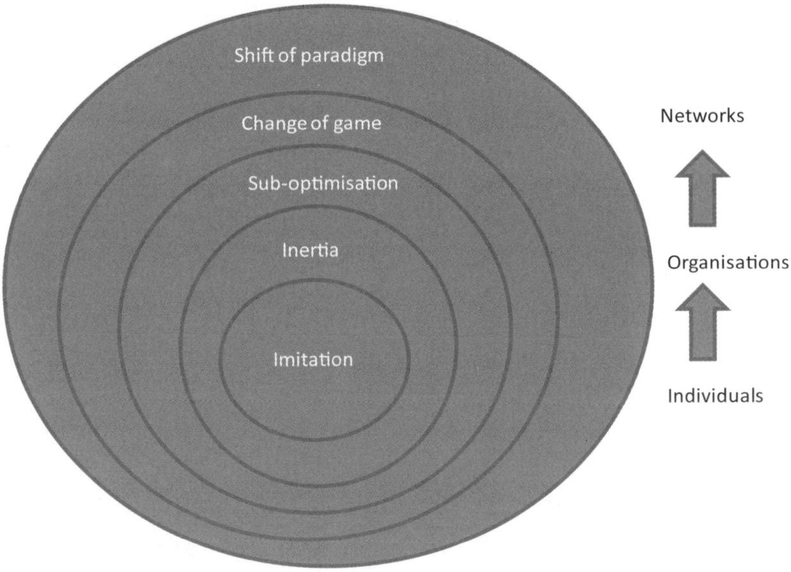


**Figure 6.** Levels of systemic changes (modified cf. Ref. [36], p. 4)

tendency to push one´s strength to its limits transforms that strength into a destructive weakness. After sub-optimisation follows change of game, which typically means transformation of the problem and the appearance of a counterproductive process that converts success to failure. The act of playing game successfully changes the game itself. A new competitive game emerges as soon as a competitive challenge is met. The nature of the problem is transformed with the change of game. The biggest possible change is a shift in a paradigm. The cumulative effects of imitation, inertia, sub-optimisation and change of the game manifest themselves in the fifth force, a shift of a paradigm[36] (pp. 4, 7–8). In the following, we will discuss the systemic change of the OI paradigm, especially from an individual and small group point-of-view while trying to predict the future of the OI paradigm.

Many scholars and practitioners agree that, to generate a single commercially successful innovation, thousands of raw ideas are needed[37] (e.g., Ref. [32]). Therefore, especially the early phase of innovation process,[38] also known as the fuzzy front end (FFE) of innovation,[39] should focus on the idea quantity. Innovation, creativity and learning are inter-related concepts[40] even though they may have varying meanings in different disciplines.[41] It has been argued many times that new knowledge and creativity always start from individual efforts, which later on can be transformed into valuable organisational knowledge to contribute to organisational innovation.[42]

We argue that it is important to understand that the source of new ideas and innovations is a co-evolution process between the individual and his/her social environment.[43] As in the case of companies, also individuals are constantly balancing between open and closed strategies during their innovation process, which in this chapter we consider as closely synonymous with the process of learning and creativity. According to Shirouzu *et al.*[44], individuals use their *inner resources*, including memory and intentions, and *outer resources*, including social and material resources for learning. In our opinion, an individual's inner resources can be aligned with the closed innovation paradigm, while his/her outer resources are aligned with the open innovation paradigm. To simplify, individuals are switching to outer resources when they discover limitations in their inner resources, the way companies do when their internal resources are exceeded. Both large and small companies can adopt the OI paradigm or some elements of it[45] (see pp. 805–806). As a result, the individual's attitudes and efforts have critical influence on the innovation process outcome.

Therefore, it is important to clarify some of the known theoretical findings which are related to creativity and learning of individuals and small groups and which can be linked to open vs. closed innovation strategy discussion. In this section, we will mainly focus on the discoveries of FFE, which typically includes stages from idea generation to decisions on further development.[46,47]

Of the different methods proposed for FFE, brainstorming[48] is among the most well known, applied and studied. Therefore, we will use it as an example to illustrate our points relating to the discussion about when to apply open vs. closed innovation strategy at the individual level. Brainstorming is a specific procedure which has a set of rules for idea generation.[49,50] These rules include: (1) Limitation of group size between five and seven, (2) prohibition of criticism, (3) encouragement for freewheeling, (4) avoidance of over-structuring, (5) note-taking during the session, (6) emphasis on quantity and variety, (7) building on the ideas of others by combining and improving them and (8), later on, editing, categorisation and selection of the best ideas for a possible implementation. It is noteworthy that both brainstorming and FFE include not only idea generation but also idea selection, the latter also known as idea screening.[57] Consequently, we should consider FFE of open vs. closed innovation strategy as a two-step process, which includes idea generation and idea selection as a follow-up step. However, the existing brainstorming and creativity literature, especially, has mainly focused on the idea generation process, while the body of knowledge focusing on idea selection process has remained clearly less represented.[52] Recent study by Hammedi *et al.*[53] proposed some guidelines for idea screening teams to improve their decision-making abilities. The guidelines aim to stimulate openness and argument-based discussion, allowing team members to stop-and-think and adapt available tools and models for specific tasks. By highlighting the importance of open discussion, these findings at least partially support the OI paradigm as a relevant approach.

Coming back to our initial assumption, derived from Osborn[48], that idea quantity breeds quality, a successful FFE innovation process should emphasise quantity. Most typically, brainstorming is implemented as an interactive group activity, although it can also be carried out as an individual process. In this paper, we will align individual brainstorming to closed innovation and interactive group brainstorming to open innovation. Interestingly, it is commonly incorrectly assumed that interactive group brainstorming is more effective than a "nominal group" approach which is also known as hybrid process.[52] In the nominal or hybrid approach, individuals perform first in isolation, with no interaction. Their productivity is later on combined into group effort.[54,55] A long stream of studies has shown that nominal groups (a.k.a. hybrid groups) outperform interactive groups in terms of number of generated ideas[56], which, as pointed out earlier, has been argued to lead a better quality. Thus, a pure "open innovation" strategy at individual or small group level might not be as effective as a hybrid strategy, which combines the close and open approaches. Evaluation apprehension, social loafing (a.k.a. free riding) and productivity blocking are typically offered as reasons for inferior performance of interactive groups.[57]

However, despite of inferior performance, interactive brainstorming (i.e., open innovation approach) remains a popular approach at individual and small team level. As a matter of fact, most individuals believe that they can generate more ideas in a group than alone, partly due to the possibility of social comparison in group brainstorming and a tendency to appropriate others' ideas.[58] Yet the common brainstorming technique of building on others' ideas is found to be counterproductive, both in terms of quantity and quality.[52] This evidently raises questions: If interactive groups (i.e., those who are adhering to open innovation) are counterproductive, how can companies which are interacting and collaborating with each other produce better innovations?

Furthermore, although nominal groups are able to generate more and better ideas in some settings, their ability to identify and select the best ideas is not always better when compared to interactive groups[50] even though some studies have found support for nominal groups' better ability to assess the quality of ideas.[50] Hence, a nominal group is not necessarily able to make use of their starting point which is better than the starting point of an interactive group. Therefore, also in the case of open vs. closed innovation approach, the ability to generate ideas might not be as important as the ability to select the best ones. This finding can have interesting interpretations on a company level. It might be a good strategy for a company, together with their OI partners, to generate or collect new ideas while reserving any further development selection decisions for themselves.

Besides affecting the interactive vs. nominal group composition, the generic structure of brainstorming process can also influence the brainstorming outcome. For example, instructions before brainstorming session focusing on the quantity of ideas can enhance productivity significantly more than a session where participants do not receive such instructions.[59] Moreover, studies comparing the quantity vs. quality focus of brainstorming instructions have produced mixed outcomes; yet some recent studies argue that quantity instruction is the most beneficial strategy for brainstorming.[58] It is noteworthy that when idea generation and selection are presented as a single task, especially for an interactive group, they generate fewer ideas than when presented as two separate tasks.[50] Therefore, guidelines and instructions related to the open innovation process should be regarded as critical success factors for the process. For a company level strategy discussion, it might be a good starting point to highlight idea quantity as a critical goal for OI strategy.

Personal and contextual characteristics are known to influence the creativity of individuals.[59] Therefore, these factors should be noted also when open innovation process is applied. It is known that, besides instructions and assessment tools, the learner's personality, cognitive style and academic abilities influence the learning results and learning process.[60–69] Therefore, when applying an open vs. closed innovation approach, it would be especially important to provide opportunities for multiple learning styles and different kinds of learners (a.k.a. innovators).[66–69] Collaborating with others might be a useful strategy for someone, while for someone else it might be a significantly weaker approach than that of innovating alone. Thus, a company with many lone innovators might not be that successful in applying the OI strategy.

A series of studies[70–72] indicate that motivation to learn, which we align in this study with creativity and innovation, depends upon a complex mix of intrinsic and extrinsic factors. Intrinsic motivation[73] — a tendency to engage in activities for their own sake, just for the pleasure derived in performing them or for the satisfaction of curiosity,[74] without any external reward or punishment[75] — is one of the most important factors related to learning, and it is also strongly linked to creativity.[76] Extrinsic motivation, on the contrary, refers to doing something because it leads to a distinctive outcome[77] such as compliance, recognition, and grades and rewards.[74] Unfortunately, all humans are not equally intrinsically motivated, and the level and the orientation of motivation most likely vary among the FFE of innovation participants in spite of example their brainstorming method. Easy-to-apply methods to measure the level of intrinsic motivation include self-reports of interest and enjoyment of the activity *per se*[77], as well as task-specific[78] or general measures such as one's intrinsic motivation for school in the case of learning.[79] If the self-reports of employees indicate higher preferences to innovating by lone individuals, then applying the OI paradigm should be questioned even if extrinsic motivation rewards are planned to be used.

As presented above, we have identified multiple factors which, from an individual innovator's point of view, are possibly influencing the success of the open innovation process.[14] Thus, when directing a discussion in favour of either the open or closed innovation approach, one should make a conscious decision related to the different options for carrying out the FFE of the process. These decisions include: (a) whether to use an interactive or a nominal/hybrid group or an individual approach (i.e., open vs. hybrid vs. closed innovation approach), (b) how to form collaborating groups, (c) when and what kinds of instructions and task orientations should be given to those who collaborate before starting the FFE of the innovation process and during it, and (d) paying attention also to the idea screening stage rather than just emphasising the importance of idea generation stage.

## 5. The Quartet Helix Model and New Dynamics of Open Innovation Processes

The Triple Helix Model is one of the best-known frameworks for describing collaboration among universities, industry and the government and explaining structural development in innovation-driven, knowledge-based economy.[80,81] In the Triple Helix Model, each actor has its own task: Universities engage in research, industries manufacture, and the government secures a level of stability for maintaining communication and interaction among those participating.

The Triple Helix Regime operates on these complex dynamics of innovation as a recursive overlay of interactions and negotiations among the three institutional spheres. The partners engage in collaboration and competition as they calibrate their strategic direction and niche positions. The Triple Helix reveals that our social world is more complex than the natural one. Over time, the following three alternative Triple Helix models have evolved (Figure 7): (1) The etatistic model, (2) the "laissez-faire" model and (3) the integrated model.[82]

According to Etzkowitz, information production has moved from universities to university–government–industry interaction, or towards Mode 2. As a result, it is argued that only the integrated model really supports innovation in knowledge societies. It is possible that the "Etatistic" and "Laissez-faire" Triple Helix models are

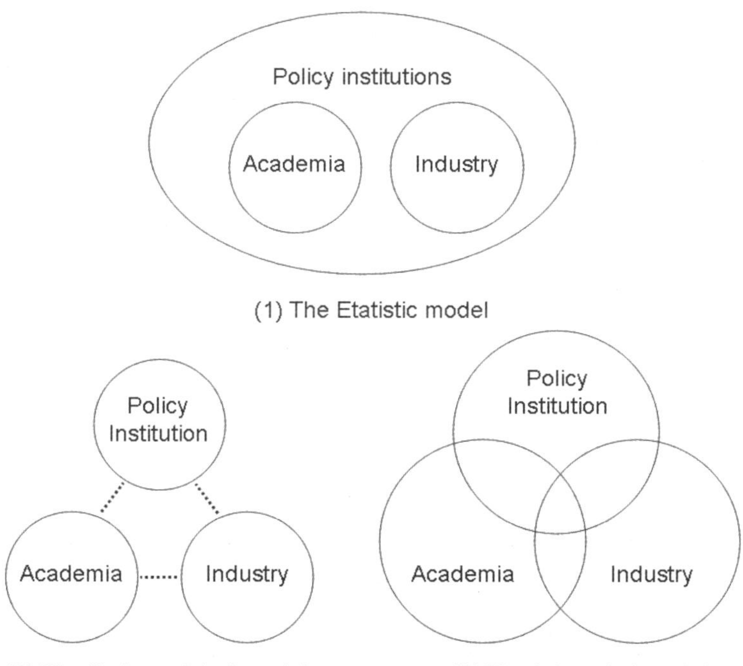


**Figure 7.** Three alternative Triple Helix models

cooperation models, which have, in fact, often actively discouraged novel innovations. Would-be innovators can become frustrated by bad management and conservative management processes built to ensure discipline, alignment and conformity rather than to provide support for creativity, innovation and experimentation.

The latest step in the Triple Helix development has been the concept of Triple Helix Systems of innovation,[83,84] which was recently introduced as an analytical framework that synthesises the key features of Triple Helix interactions into an 'innovation system' format, defined according to the systems theory as a set of components, relationships and functions. In this new format, a novel distinction has been made between: (1) R&D and non-R&D innovators; (2) "single-sphere" and "multi-sphere" (hybrid) institutions; and (3) individual and institutional innovators. The new strategic relationships between these components are synthesised into five main types of operations:

(1) Technology transfer,
(2) collaboration and conflict moderation,
(3) collaborative leadership,
(4) substitution, and
(5) networking.

These new elements of relationships help innovation managers to understand, more deeply, the dynamics of the Triple Helix, especially because most postmodern societies have moved from industrial logic to knowledge society logic. Today, knowledge networks are playing much bigger role in innovation processes than before. Conventional industries in many innovation processes are substituted by value networks and knowledge brokers.[83,84]

To the Triple Helix Model, we can add one important element: Customers and end users. This version of innovation system can be referred to as the Quartet Helix Model. We can construct a futures table where we have linked the open innovation paradigm to the Quartet Helix Model.

This scenario table (Table 4) tells us that there are different kinds of innovation management models. In the Quartet Helix, nations

**Table 4.** Open innovation paradigm and the Quartet Helix Model

| | **Innovation model 1: Closed IM** | **Innovation model 2: Symbiosis IM** | **Innovation model 3: Open IM** |
|---|---|---|---|
| Government | Closed innovation paradigm dominates | In between closed and open innovation model | Open innovation paradigm dominate |
| Academia | Closed innovation paradigm dominates | In between closed and open innovation model | Open innovation paradigm dominate |
| Industries | Closed innovation paradigm dominates | In between closed and open innovation model | Open innovation paradigm dominate |
| Customers | Closed innovation paradigm dominates | In between closed and open innovation model | Open innovation paradigm dominate |

can create many alternative combinations of open innovation management. The Integrated Triple Helix Model is probably more favourable for open innovation than the Etatistic and "Laissez-faire" models. National innovation models are either totally open or totally closed. We present here a new theoretical concept, the symbiosis innovation model, which suggests that the closed innovation and closed innovation models can "live in a symbiosis". This is an evolutionary interpretation of OI.

A policy choice where the open innovation paradigm dominates in all sub-systems of the Quartet Helix Model is an extreme form of innovation management. Another extreme form of innovation management is a policy choice where the closed innovation paradigm dominates in all sub-systems of the Quartet Helix Model. Of course, in middle-of-the-road innovation management, all the sub-systems of the Quartet Helix Model have adopted an approach between the closed and the open innovation paradigm.

The Quartet Helix Model, which goes beyond the Triple Helix Model, is visualised in Figure 8. Its fourth element is "Customers and end-users". Thus, Quartet Helix includes also the idea of consumer-driven innovations or user-driven innovation.

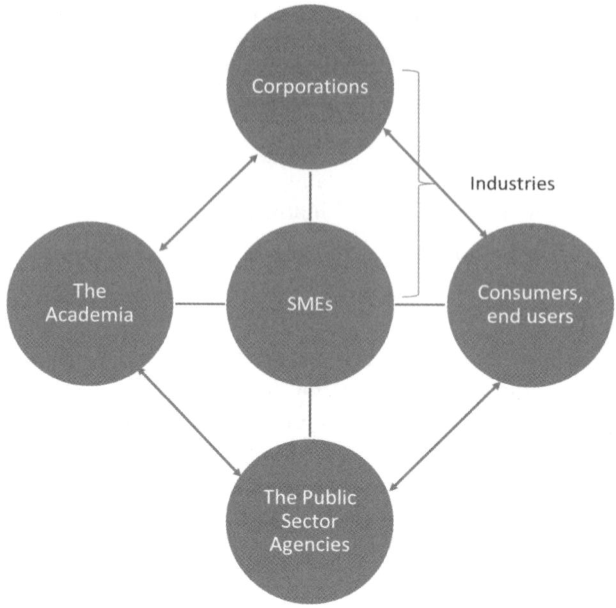


**Figure 8.** Quartet Helix Model of innovation (4 key stakeholders)

## 6. Open Innovation Management: Roadmaps to the Future

In Figure 9, we visualise an integrated open innovation management model with product markets and technologies. This model includes analyses of product markets, technologies and technology markets. This kind of model can be very useful in planning future steps of product development and needed technologies. It helps firms to pay attention to technology markets they rely on.

Recent OI research is closely linked to the analysis of collaboration and to interoperability of systems relevant for firms. Figure 10 visualises an Enterprise 3.0 Framework. The figure also reveals some key tendencies of the future of the OI paradigm. We can expect that the Semantic Web and the emerging Web 3.0 will increase firms' collaboration possibilities. This also implies a new era for open innovation management.[31]

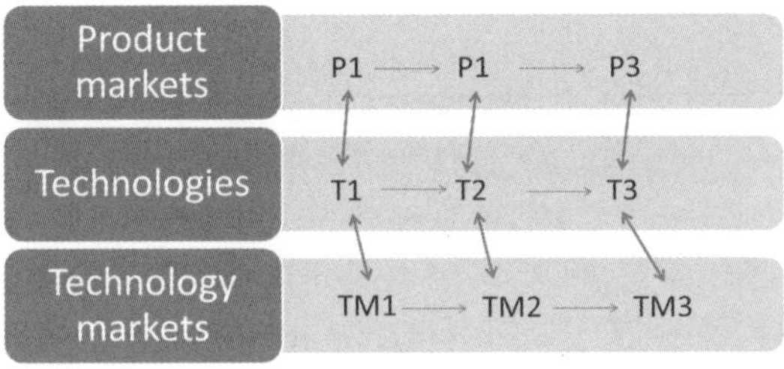


**Figure 9.** Integrated open innovation management model

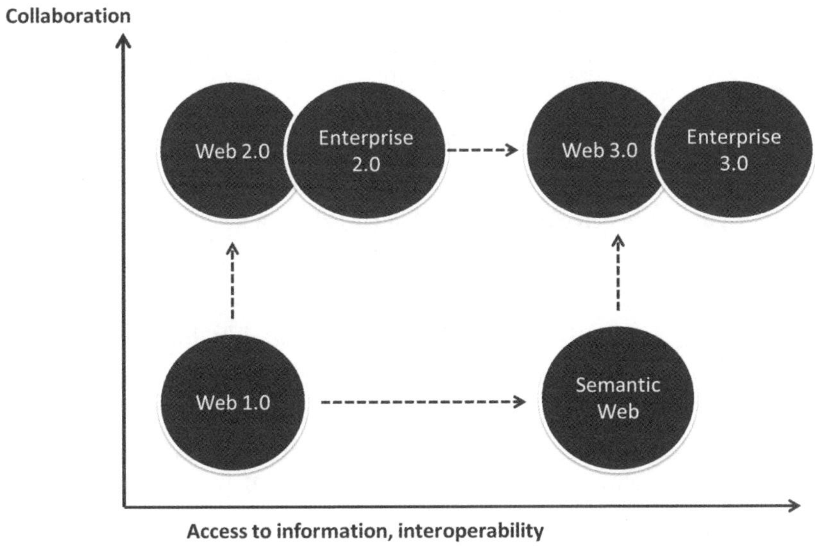


**Figure 10.** Scenarios of enterprise in different collaboration and interoperability environments[83] (See p. 8932)

The future of the OI paradigm depends on the development of Web 2.0 and Web 3.0. Especially the Semantic Web will increase access to information and interoperability for firms' and other agencies' systems. As we can observe in Figure 10, the scale of collaboration and interoperability will increase in the future. This also implies that there will be better technical and economic opportunities for companies that use OI management tools.

In Figure 11, we have outlined baseline scenarios for the percentages of population electronically networked for years 2010–2050.

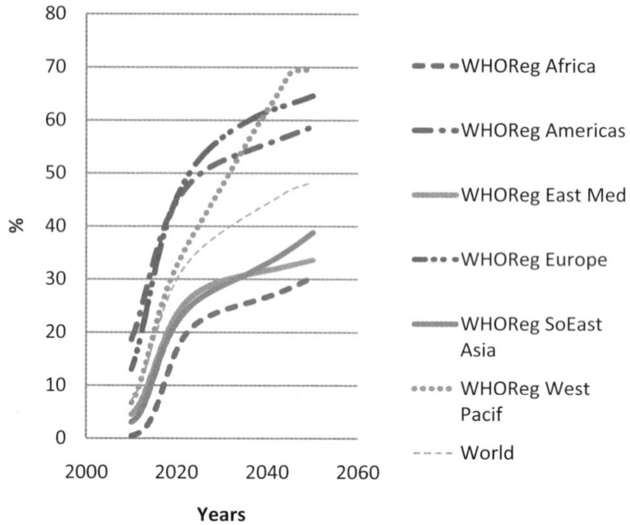


**Figure 11.** Percent of population (%) electronically network[86]

This analysis, which is based on the IFs model,[85] indicates that the process of networking is the strongest in the West Pacific Region, in Europe and in the Americas. Lower level of networking is taking place in the developing world in Africa, South East Africa and in the Middle East. In the entire world, the share will be less than 50% in 2050, according to the baseline scenario.

In Figure 12, we have visualised the Freedom Index for world regions. This baseline scenario indicates stable progress in the freedom of people.

These two analyses indicate good conditions for the OI paradigm, because freedom and networking are expected to develop further in the world's regions. Of course, there are considerable regional differences in favourable conditions, as Figures 11 and 12 indicate. Compared to China, India and the developing world, there will be more freedom among citizens of Europe and the US. The conditions for the OI paradigm will probably be better in these Western countries due to their more highly networked and free

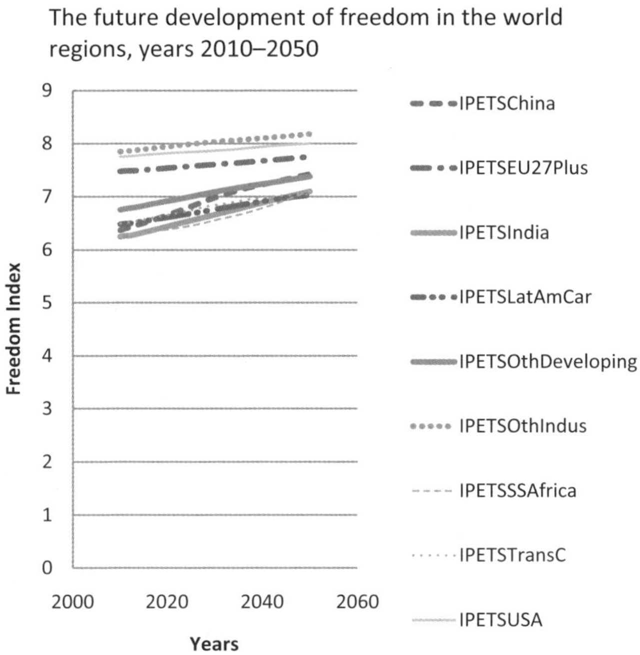


**Figure 12.** Freedom index of the world[86]

business environments. Thus, if we analyse networking and freedoms as the key drivers of OI, we can expect that there will be considerable differences in the preconditions for OI management. This is good to remember when we analyse the futures of the OI paradigm.

## 7. Alternative Innovation Models of Private Enterprises, with Special Attention to Foresight–Innovation Linkages

Originally, the so-called three-phase model of R&D, production and marketing was presented by Takeuchi and Nonaka.[87] Later, Kaivooja[88] presented a more complex interaction model, which includes both innovation management and foresight process. In this section, we shall present alternative innovation models for private enterprises, with special attention to foresight–innovation linkages. Innovation model here refers both to the open and closed innovation management models.[88] These models will probably be used in the futures of innovation management.

These systemic interaction models are based partly on an empirical study which was presented by Leiponen and Drejer.[89] They identified five behavioural/technological regimes in the Nordic countries: (1) Supplier-dominated regime, (2) production-dominated regime, (3) science- or scale-based regime, (4) market-driven regime and (5) passive or weak regime. Most industries and firms in the Nordic countries have these regimes represented. Firms do not have homogenous foresight or innovation management styles or "perfect foresight" as many economists routinely tend to and want to assume in conventional economic theory. The differences between market push and market pull alone make their innovation management models different.

The models of interaction between the foresight system and the innovation process are presented in Figures 13–19. There is actually a very relevant aspect of innovation management, the foresight aspect, which is related to innovation process in many alternative ways.

In Figure 13, the so-called IFO model is presented. The IFO model is based on the results of innovation process. Typically, in the IFO process, foresight analyses serve commercialisation activities in production and marketing. The IFO model is based on the technology push mechanism.

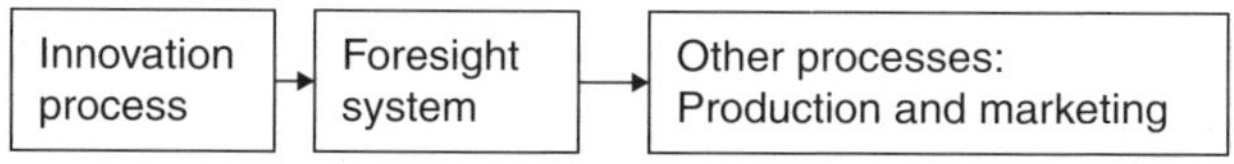


**Figure 13.** Model I: Innovation–Foresight–Other Processes (IFO) model

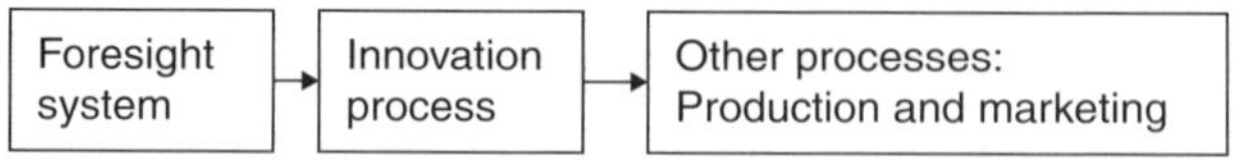


**Figure 14.** Model II: Foresight–Innovation–Other Processes (FIO) model

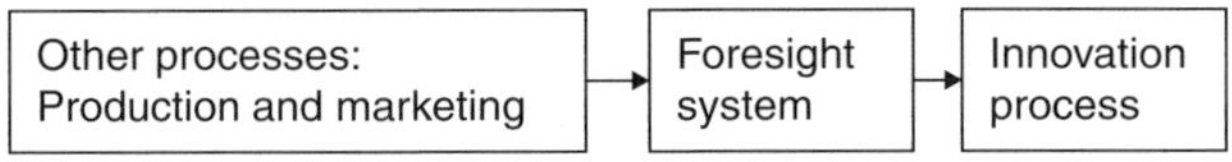


**Figure 15.** Model III: Other Industrial Processes–Foresight–Innovation (OFI) model

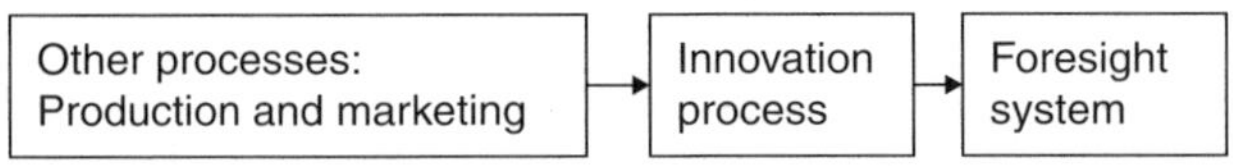


**Figure 16.** Model IV: Other Industrial Processes–Innovation–Foresight (OIF) model

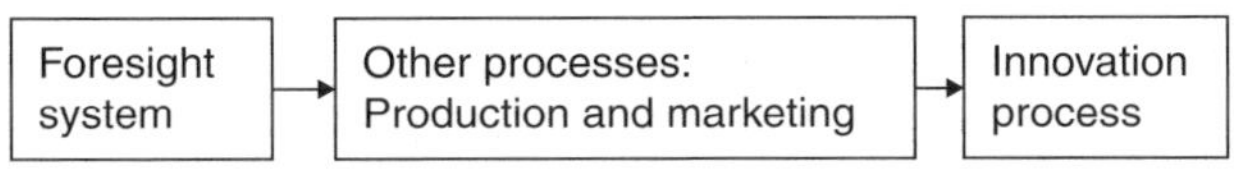


**Figure 17.** Model V: Foresight–Other Industrial and Marketing Processes–Innovation (FOI) model

Figure 14 shows the FIO model, where foresight systems play a catalytic role in relation to innovation process. Typical elements of foresight analyses, including trend analyses, scenario analyses and Weak Signal and Wild Card (WI–WE) analyses, are used to stimulate innovation thinking and innovation process. The FIO model is mostly based on the technology push mechanism.

The OFI model shown in Figure 15 is based on firm-level processes and a foresight system. In the OFI model, innovation processes are secondary processes to a firm's production and marketing processes and foresight processes. The OFI model is mostly based on the technology pull mechanism.

In Figure 16, the OIF model is presented. The OIF model is mostly based on the technology pull mechanism. In this model, production and marketing units of a company help innovation units to arise; after that, foresight activities make other analyses relevant for a new innovation.

Figure 17 depicts the FOI model, which is based on the technology pull mechanism. In this model, foresight analyses are first used in production and marketing processes. These analyses then lead to new innovations.

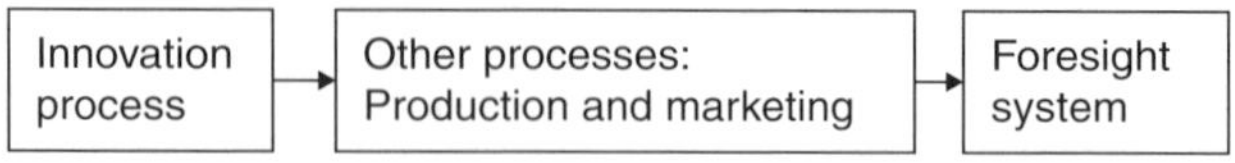


**Figure 18.** Model VI: Innovation–Other Industrial Processes–Foresight (IOF)

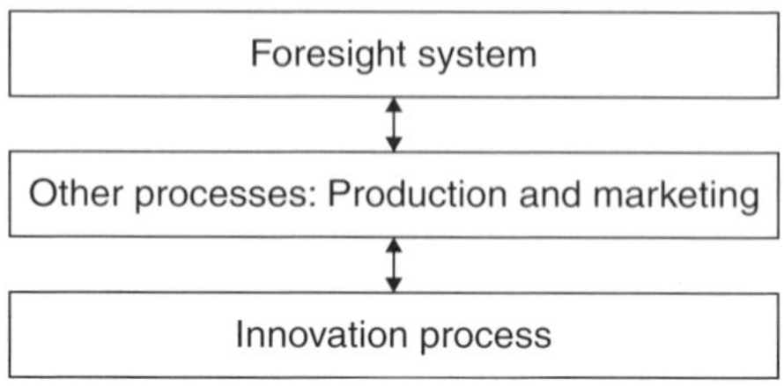


**Figure 19.** Model VII: Interactive Simulative Process (ISP) model

In Figure 18, the IOF model is shown. The IOF model is based mostly on the technology push mechanism. The starting idea of the IOF model is based on other production and marketing processes in innovation management and use foresight in the final phase of commercialisation.

Finally, the ISP model is presented in Figure 19. The fundamental idea of the ISP model is to perform three processes simultaneously. This ISP model is organisationally very demanding for companies and corporations.

These models can be linked to empirical realities of companies and corporations. Kaivo-oja[88] (p. 25) presented Table 5 in his theoretical study of innovation management in firms.

Thus, we can conclude that the role of foresight knowledge is different in different systemic interaction models. Foresight knowledge can be secret (closed innovation paradigm) or open to everyone (open innovation paradigm).

In recent foresight and futures studies, van der Duin *et al.*[28] (pp. 65–66) presented a cyclic innovation model which is based on an idea quite similar to the idea behind the ISP-model. This CI-model is shown in Figure 20.

The model is interesting, because being a "process model" it can be linked to systemic interaction models presented in Figure 20. This model also underlines the need to integrate foresight and innovation

**Table 5.** The role of foresight knowledge in different systemic interaction models

| Systemic interaction models | The role of foresight knowledge |
|---|---|
| IFO-model | To promote innovation realisation in production and marketing |
| FIO-model | To catalyse, first the innovation process and then the production and marketing processes |
| OFI-model | To analyse current production and marketing processes and make new innovations on the basis of foresight analyses |
| OIF-model | To create and carry out foresight analyses on the basis of production and marketing activities and create innovations |
| FOI-model | To catalyse, first, the production and marketing analyses and operations and then produce novel innovations |
| IOF-model | To create and carry out foresight analyses on the basis of novel innovations and related new production and marketing processes |
| ISP-model | To implement three system interaction processes simultaneously, leading to better and integrated foresight analyses and new innovations in production and marketing units |

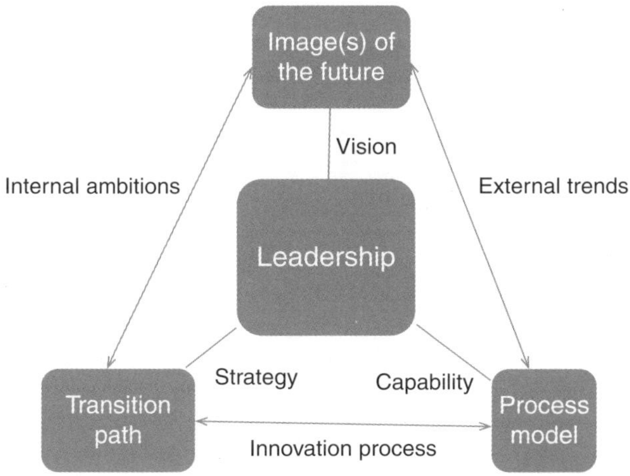


**Figure 20.** Cyclic innovation model[90] (p. 66)

activities. According to Rohrbeck and Gemünden[90], the three main roles of foresight are those of: (1) Initiator, (2) strategist and (3) opponent. The role of foresight always has impacts on innovation process and management, depending on whether the foresight has the role of the initiator, that of the strategist or of the opponent. These different roles may help the firm to innovate and remain at the competitive edge.

Table 6 provides a comprehensive summary of the scope of foresight activities. As we can see, foresight activities can be either open

**Table 6.** The scope of the foresight activities and key roles in innovation processes[106,107]

| Activity | Initiator role | Strategist role | Opponent role | Scope |
|---|---|---|---|---|
| Inspirational workshops | Secondary role | Primary role | | Contract partners |
| Business modelling | Secondary role | Primary role | | Contract partners |
| Business case analyses | | Primary role | Secondary role | Contract partners |
| Series of future workshops | Primary role | Secondary role | | Closed network |
| Future studies | Primary role | Secondary role | | Open (organisations and end-users) |
| Business field exploration | Secondary role | Primary role | | |
| Thematic innovation radar | Primary role | | | |
| Working groups | Primary role | Secondary role | Secondary role | |
| Business modelling | | Primary role | | |
| Business case analysis | | Primary role | Secondary role | |
| Networking on demand | | Primary role | | |

(*Continued*)

Table 6. (*Continued*)

| Activity | Initiator role | Strategist role | Opponent role | Scope |
|---|---|---|---|---|
| Action lines | Secondary role | Primary role | | Closed network |
| Experience and living labs | | Primary role | Secondary role | Open (organisations and end-users) |
| Testbeds and simulation tools | | | Primary role | Closed network |
| Spearhead research | | Primary role | Secondary role | Closed network |
| Business modelling | | Primary role | Secondary role | Open (organisations) |
| Technology transfer programs | | Primary role | Secondary role | Open (organisations) |
| Innovation radar | Primary role | Secondary role | | Open (organisations) |
| Annual selection process | | Secondary role | Primary role | Closed network |
| Best-practice benchmarking | | | Primary role | Open (organisations) |
| Business developer program | | Primary role | Secondary role | Closed network |

*Source*: Refs. [3] (pp: 62–78) and [90].

or closed, depending on the scope of the project. Also the roles in activities can be different (primary of secondary).

Table 6 is a very good example of the symbiotic relationship of OI and CI. These approaches are used together, and obviously they create mutual synergies for companies.

Many companies and firms are interested in future roadmaps. There can be separate future roadmaps for markets, services, products and technologies. Figure 21 presents a very basic skeleton model of futures roadmap. This kind of roadmap can include both closed innovations and open innovations relevant for products and services.

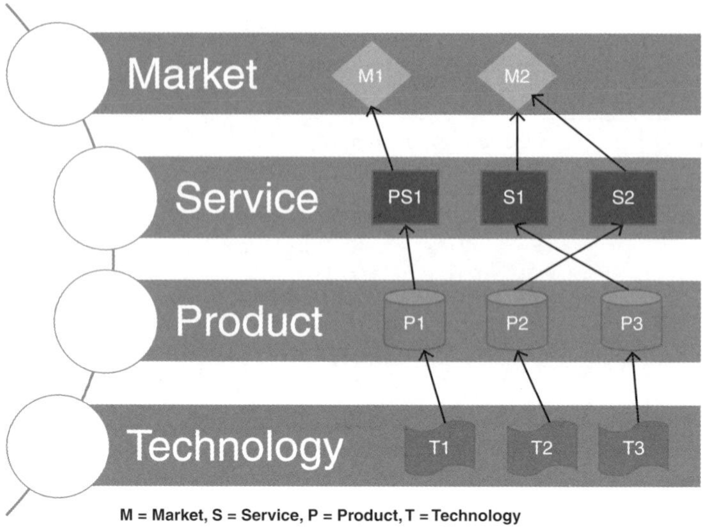


**Figure 21.** Open innovation thinking and a future roadmap model with market, service, product and technology analyses

# 8. Radical Innovation Management and Open Innovation Paradigm

In many organisations, people need a license to think (out-of-box-thinking) and act (out-of-box acting) in a different way. Innovating futures and radical innovations is going to be a key research field in OI management. In the field of futures studies, typical methods have been (1) brainstorming workshops, (2) strategic serendipity processes, (3) qualitative and quantitative scenario analyses, (4) weak signal/wild cards analyses, (5) innovation tournaments, (6) innovation competitions, and (7) innovation scorecard analyses.

A typical approach has been to raise ambition and competition to create really radical breakthrough innovations. Because incremental innovations are focused on improving standards under current market conditions, they require more stable and conventional innovation management approaches. Radical innovations are doing new things, which require new innovation management approaches.

Typically, these kinds of future inventions include (1) idea generation, (2) idea development and (3) a testing or prototyping phase.

In recent years, there has been a lot of discussion about how we can find radical pathways to innovation management. Two interesting approaches include science-fiction-based innovation[91] and nostalgia- and vintage-based innovation.[92] In this field, the main idea is to take distance to current reality and find something radical on the basis of historical or future time difference.

In today's business environment of rapid technological change, companies cannot rely on incremental innovation alone. To sustain long-term competitiveness, companies need to generate radical innovations as well. Modern innovation literature possess multiple definitions for innovation, yet most commonly the term innovation can be related to something new[93,94], which has been put into practice[95] while bringing value to customers and organisations.[96,97]

The classification of innovation itself as either incremental or radical[98] is not without its scientific critics[99], and several overlapping terms and definitions, as a close equivalent of radical innovation, have been proposed. Among these are *disruptive*[100], *revolutionary*[101], *discontinuous*[102], *pioneering*[103] and *breakthrough*[104] *innovation*. Typically, experts and media commentators tend to recognise a radical innovation when they see it, but it is exceptionally difficult to define radical innovations *ex ante*.[105]

In the field of innovation management, the challenge of radical innovation management has received a great deal academic attention in recent years. It is widely understood that large corporations are capable of creating and commercialising new products and services than offer incremental benefits to the market. Established companies, especially, struggle in their attempts to develop new radical breakthrough innovations that would confer wholly new business and rejuvenation opportunities. Many empirical studies imply that radical breakthrough innovations are linked to superior performance.[106,107] On the other hand, established organisations experience path dependency.[108,109] Many of them are ensnared in what is familiar to them, in what they have always done. Too often firms' futures are direct extensions of their past. That is why their new ideas, inventions and innovations tend to be quite similar to their present ideas and inventions. That is why there are very good reasons for established organisations to turn to the OI approach for refreshment.

In a way, the OI paradigm can help established organisations to solve the classical path dependency problem. Actually, escaping path dependency may require the use of open innovation management tools and techniques.

Established organisations are also reliant on established routines, which promote efficiency and automation over creativity and flexibility. In many organisations, core capabilities become core rigidities.[110] Because of this fundamental capability problem many organisations have started to develop strategies for dynamic capabilities. It is not easy to combine operational excellence with strong innovation culture in the highly competitive environment in which companies operate. Reliance on established routines reduces an organisation's ability to cope with uncertainty in any manner other than by ignoring it or by electing not to engage in opportunities that would expose them to it. This kind of process precludes most opportunities that could result in radical new breakthrough innovations, since every radical breakthrough innovation, by definition, is unique. Situation-specific knowledge brokerage must necessarily be developed, but many routinely working organisations do not have time for these kinds of critical tasks. Some companies are more prepared than others to act in the face of *ex ante* uncertainty.

We can sum up by stating that organisations experience path dependency and engage in uncertainty avoidance, which weakens their possibilities to introduce radical innovations. Creating radically new paths is not an easy task for most companies. Companies must introduce better risk management approaches and innovation management tools which allow better management of radical innovations. O'Connor *et al.*[111] (pp. 80–81) conclude that modern companies need two critical elements to manage radical innovation: (1) Options mentality and (2) experimental process.

Options mentality, as a risk management by learning practice, is connected to harvesting strategy. Concerning harvesting strategy, one variable is critical: legitimacy. According to empirical results, legitimacy is one of the initial hurdles that an RI Portfolio manager would need to overcome. Experimental processes are linked to options orientation as a risk management by learning practice. Project and portfolio evaluation and ongoing oversight using option orientation are strongly, directly related to RI success.

These results are highly relevant for open innovation management. They indicate that OI management can have very positive role in strengthening RI management by (1) helping innovation management process to tackle legitimacy biases and problems and by (2) paying critical attention to the diversity of project portfolio under oversight. These two broad issues (legitimacy and diversity of R&D project portfolios) are surely requiring more empirical research activity in the future.

Especially issues like alliance portfolio diversity and specification of alliance portfolio capability are key issues for radical innovation management. Recent innovation research indicates that there is an inverted-U relationship between incremental and radical innovation outcomes in relation to alliance portfolio diversity.[112] The shape of the U-curve relationship has also implications for open innovation management. The inverted-U relationship indicates that there is a certain critical point where increasing alliance portfolio diversity leads to worse outcomes of innovation management both in incremental and radical innovations. Thus, too many open innovation partnerships can be harmful for successful radical innovation and incremental innovation developments.

There may be motivational reasons behind the inverted-U curve relationship. The reward system must be extensive if there are many partners in the innovation process. Keeping a large consortium of teams motivated for radical innovation process requires a lot of motivation and rewards. The reward system should be designed to foster extrinsic autonomous motivation if an employee is not intrinsically motivated[113] (p. 147). The human side of open innovation management in radical innovation process needs more scientific attention.[114]

New results reveal that companies that emphasise inside-out open innovation are more likely to create radical innovations and tend to sell a greater number of new products. On the other hand, companies pursuing closed innovation are more likely to exhibit higher incremental product innovation performance.[94]

Still, one big scientific challenge is to understand how the Open Innovation paradigm could be linked to radical innovation management. Contrary to the case study results of Inauen and Schenker-Wicki[95], many scholars have noted that the OI paradigm will be more useful in the creation of incremental innovations rather than in the creation of radical innovations. We propose that the OI paradigm can produce and promote both incremental and radical innovation. The methods used in the innovation management can produce different results, depending on perceptions and experiences of the respondents in empirical case studies.

We know that radical innovations are unevenly distributed through time. Radical innovations are often the drivers of incremental innovations. Radical innovation typically occurs with new technologies and in new markets. New technologies are typically untried in practice, lack critical value chain elements, provide limited product advantages, and lack supporting business models. Radical innovations often create totally new classes of products or services.

Outstanding radical innovations form a precondition for incremental innovations. If the results of Inauen and Schenker-Wicki[94] could be generalised to innovation management, we could claim that open innovation management is needed to create radical innovations and closed innovation management guarantees that the next wave of incremental innovations follow from some important radical innovations. Another important result of radical innovation management is that the user involvement of the "right users" at the "right time" in the "right form" works for the development of radical innovation management. Pure crowdsourcing strategy does not easily lead to radical innovations[115]; more probably it leads to incremental innovations.

How to develop and commercialise radical innovations in a better way? On the basis of current scientific literature, we can answer that the necessary elements of radical innovation management are

(1) an identifiable organisation structure responsible for radical innovation, (2) leadership and culture associated with a learning orientation, (3) appropriate process for high uncertainty and high risk domain, (4) skills and talent development for innovation, (5) governance and decision making specific to innovation and (6) appropriate metrics of innovation process[111] (pp. 71–72).

Radical innovations require a system which can be sustained over time. In many studies, one key variable in the radical innovation process has been identified. There are two typical firm-level strategic orientations towards radical innovations: Technology orientation and customer relationship orientation. Radical elements can be identified in (1) technological radicalness, (2) business model radicalness and (3) market radicalness. Recent empirical studies indicate that there are cause–effect relationships between orientations and dimensions of radicalness (Table 7).

Thus, both types of strategic orientation play a major role in fostering radical innovations on all dimensions. This means that in order to make effective open radical innovation management operations

**Table 7.** The effect of orientations and moderating roles of market uncertainty on technological, market and business model radicalness (p. 597)

| | Technological radicalness | Market radicalness | Business model radicalness |
|---|---|---|---|
| Effect of technology orientation | Positive | Positive | Positive |
| Effect of customer relationship orientation | Positive | Non-significant | Positive |
| Moderating role of market uncertainty on the effect of technology orientation | Negative moderation | Negative moderation | Non-significant |
| Moderating role of market uncertainty on the effect of customer relationship orientation | Non-significant | Non-significant | Non-significant |

*Source*: Ref. [116], (p. 597).

both technology and customer relationship orientations must include the use of open innovation tools and approaches, which increases technological radicalness, market radicalness and business model radicalness. Thus, firms using the OI approach in developing novel radical innovations must evaluate potentially radical innovations subjectively from the technological, market and business model perspectives. The role of uncertainty is crucial. We can conclude that managers should consider environmental contingencies. From Table 7, we can note that it may not be rational to push all types of radical innovation in conditions of very high market uncertainty, as the result may not be optimal. If the level of turbulence and uncertainty is very low, the opposite conclusion will be optimal. Anyhow, market uncertainty has to be taken into account in radical innovation management.

Successful knowledge transfers between systems and organisations are needed to create radical innovations. Too often organisations create closed silos and boundaries, which hinder knowledge transfers. We also know that knowledge is closely linked to the learning processes of organisations. Thus, knowledge transfers, learning culture and dynamic capabilities are key constituents of successful radical innovations. Also two other issues mentioned above, legitimacy aspects and diversity of R&D portfolio, are very important and influential for the profiting from RI management.

Obviously, we should pay more attention to open innovation management tools, which help companies to introduce more radical innovations, instead of incremental innovations, which do not provide much long-term gains in competition. This does not mean that incremental innovations are useless. In some small markets and market segments, firms might need only incremental innovations to survive, but nowadays these kinds of closed markets without a need for radical innovations are exceptions.

Radical innovations are critical to the long-term success of firms. We can conclude that radical innovations are innovations that provide the biggest boost to the company's competitive edge, in terms of technology, market position and customer value added, in general. Open innovation management can help firms to boost these key variables in many ways. Issues like project OI portfolio management, customer-driven orientation in OI management, technology orientation in OI management, risk management in OI management and dynamic capabilities in OI management, among many other issues, need more research activity in the field of open innovation management.

Probably in the future there will be more discussion about radical open innovation management.

## 9. Summary

Innovation researchers have understood since the 1970s that the sources of innovative ideas and new inventions come from outside the company. Also the role of consumers is important in many innovation processes. Business model innovations are today a part of open innovation thinking. The role of external knowledge and ideas is widely acknowledged in many studies, especially in the OI research tradition. Empirically, the critical role of OI can be verified in many ways.

In this chapter, we have outlined some futures-oriented analyses of the OI paradigm. There are some key issues we would like to underline.

Firstly, the basic idea of open innovation assumes that corporate innovation activities are more like an open system than the traditional vertically integrated model. Under open innovation, large firms do not abandon the latter approach, but rather augment their traditional R&D practices within bound sourcing of external technologies throughout the product development process and use controlled outflows of internal technologies to seek new markets through outbound licensing[1,2] (see pp. 2–3). This point is relevant for business model innovations and customer-driven innovations, too.

The development of modern IO thinking is based on observing a handful of large innovative companies and their deviations from traditional practice. Our approach is somewhat critical, emphasising the role of individuals. We claim that the role of individuals should be discussed more in the context of OI management. There are still hierarchies and phenomena of vertical integration in the global markets. Multinational companies have a fair amount of internal business and collaboration. According to Chesbrough, "Under open innovation, large firms do not abandon this approach, but rather augment their traditional R&D practices within bound sourcing of external technologies throughout the product development process, as well as controlled outflows of internal technologies seeking new markets through outbound licensing"[2] (see pp. 2–3). In empirical reality of global economy, pure forms of OI or CI are not observed. This is the fundamental reason why we have presented in this chapter the concept of *symbiotic innovation management model*, which falls "somewhere between" pure OI and pure CI, where OI needs the CI approach and CI needs the OI approach.

As we have noted in this chapter, similar kind of relationship can be identified between incremental innovation and radical innovation.

Secondly, the open innovation paradigm and the closed innovation paradigm are complementary approaches. Each paradigm depends on the other. If firms prefer more OI, it takes activity from CI, and vice versa.

Thirdly, global networking is a key driver for OI activities. The functioning of networks plays a critical role in OI activities. The role of an individual and their motivation cannot be underestimated in organisations and networks. If organisations and networks are able to provide motivational incentives for individuals, there will be growing interest on OI by individuals. In other case, individuals will turn to more closed innovation processes.

Fourthly, the future Web 2.0 and Web. 3.0 are having impacts on future organisation of firms. The turbulence levels of technological change and global markets play a crucial role for firms' futures and profit functions. To put it simply: If the IO approach helps companies to profit from innovation, IO approach will be used.

Fifthly, radical innovations are a key issue for the future growth of companies. There is very obvious need to integrate theories of radical innovation to the modern IO paradigm.

## References


1. Chesbrough, H. (2003). *Open Innovation: The New Imperative for Creating and Profiting from Technology* (Harvard Business School Press: Harvard).
2. Chesbrough, H., Vanhaverbeke, W. and West, J. (2006). *Open Innovation: Researching a New Paradigm* (Oxford University Press, Oxford).
3. van der Duin, P., Heger, T. and Schlsinger, M. D. (2014). Toward networked foresight? Exploring the use of futures research in innovation networks. *Futures*, 59: 62–78.
4. Colombo, M. G., Piva, E. and Rossi-Lamastra, C. (2014). Open innovation and within-industry diversification in small and medium enterprises: The case of open source software firms. *Research Policy*, 43: 891–901.
5. Kessler, E. H. and Chakbrabarti, A. K. (1996). Innovation speed: A conceptual model of context, antecedents, and outcome. *Academy of Management Review*, 21: 1143–1191.
6. Kumar, S. and Eickhoff, J. H. (2005). Outsourcing: When and how should it be done? *Information Knowledge Systems Management*, 5: 245–259.
7. Ahn, S. (1995). A new program in cooperative research between academia and industry in Korea involving Centers of Excellence. *Technovation*, 15: 241–257.
8. Gassman, O. and Enkel, E. (2004). Theory towards of open innovation: Three core process archetypes. *In R&D Management Conference* (RADMA) (Lisbon, Portugal).
9. Cowan, R., Jonard, N. and Zimmerman, J. B. (2007). Bilateral collaboration and the emergence of innovation networks. *Management Science*, 53: 1051–1067.
10. Gassman, O., Enkel, H. and Chesbrough, H. (2010). The future of open innovation. *R&D Management*, 40: 213–221.
11. Gergils H. (2005). *Dynamic Innovation Systems in the Nordic Countries? A Summary Analysis and Assessment* (VINNOVA: Sweden).
12. Gassmann, O. and Enkel, E. (2014). Constituents of open Innovation: three core process archetypes. *R&D Management*, Forthcoming.
13. Gassman, O. (2006). Opening up the innovation process: Towards an agenda. *R&D Management*, 36: 223–229.
14. West, J., Salter, A. Vanhaverbeke, W. and Chesbrough, H. (2014). Open innovation: The next decade. *Special Issue on Open Innovation: New Insights and Evidence. Research Policy*, 43: 805–811.
15. Freeman, C. (1994). Innovation and growth. In Dodgson, M. and Rothwell, R. (eds.), *Handbook of Industrial Innovation* (Aldershot, Elgar), pp. 78–93.
16. Chandler, A. D. (1990). *Scale and Scope* (Cambridge, MA, The Belknap Press of Harvard University Press).

17. Mowery, D. (1990). The development of industrial research in U.S. manufacturing. *American Economic Review*, 80(2): 345–349.
18. Pavitt, K. (1991). What makes basic research economically useful? *Research Policy*, 20(2): 109–119.
19. Castellacci, F. (2008). Innovation and the competitiveness of industries: Comparing the mainstream and the evolutionary approaches. *Technological Forecasting and Social Change*, 75(7): 984–1006.
20. Nelson, R. R. (ed.). (1962). *The Rate and Direction of Inventive Activity* (Princeton University Press, Princeton, NJ).
21. Nelson, R. R. (1968). A diffusion model of international productivity differences in manufacturing industry. *American Economic Review*, 58(5): 1219–1248.
22. Nelson, R. R. (1981). Research on productivity growth and productivity differences: Dead ends and new departures. *Journal of Economic Literature*, 19(3): 1029–1064.
23. Dosi, G. and Nelson, R. R. (2010). Technical change and industrial dynamics as evolutionary processes. *Handbook of the Economics of Innovation*, 1(2010): 51–127.
24. van der Duin, P., Ortt, R. J. and Kok, M. (2006). The cyclic innovation model: A new challenge for a regional approach to innovation systems. *European Planning Studies*, 15: 195–538.
25. Daheim, C. and Uerz, G. (2008). Corporate foresight in Europe: From trend based logics to open foresight. *Technology Analysis & Strategic Management*, 20(3): 321–336.
26. Sorenson, O. and Fleming, L. (2004). Science and the diffusion of knowledge. *Research Policy*, 33(10): 1615–1634.
27. See e.g. Franzoni, C. and Sauermann, H. (2014). Crowd science: The organization of scientific research in open collaborative projects. *Research Policy*, 43(1): 1–20.
28. Henkel, J., Schöberl, S. and Alexy, O. (2014). The emergence of openness: How and why firms adopt selective revealing in open innovation. *Research Policy*, 43(5): 879–890.
29. Franzoni, C. and Sauermann, H. (2014). Crowd science: The organization of scientific research in open collaborative projects. *Research Policy*, 43(1): 1–20.
30. Felin, T. and Zengler, T. R. (2014). Closed or open innovation? Problem solving and the governance choice. *Research Policy*, 43: 914–925. See especially pp. 917–919.
31. Lee, S., Park, G., Yoon, B. and Park, J. (2010). Open innovation in SMEs — An intermediated network model. *Research Policy*, 39(2): 290–300.
32. Carbone, F., Contreras, J. J., Hernãndez, Z. and Gomez-Perez, M. (2012). Open innovation in an Enterprise 3.0 framework. *Expert Systems with Applications*, 39(10): 8929–8939.

33. Hung, K.-P. and Chou, C. (2013). The impact of open innovation on firm performance: The moderating effects of internal R&D and environmental turbulence. *Technovation*, 33(10–11): 368–380.
34. Chesbrough, H. and Rosenbloom, R. S. (2002). The role of the business model in capturing value from innovation: Evidence from Xerox Corporation's technology spin-off companies. *Industrial and Corporate Change*, 11(3): 529–555.
35. Felin, T. and Zengler, T. R. (2014). Closed or open innovation? Problem solving and the governance choice. *Research Policy*, 43: 914–925 (especially p. 918).
36. Gharajedaghi, J. (2006). Systems Thinking. *Managing Chaos and Complexity: A Platform for Designing Business Architecture*. Amsterdam: Butterworth-Heinemann, pp. 7–8 (especially p. 4).
37. Stevens, G. A. and Burley, J. (1997). 3000 raw ideas equals 1 commercial success! *Research Technology Management*, 40(3): 16–27.
38. Cooper, R. G. (1988). Predevelopment activities determine new product success. *Industrial Marketing Management*, 17(2): 237–248.
39. Smith, P. G. and Reinertsen, D. G. (1991). *Developing Products in Half the Time*. New York, van Nostrand Reinhold.
40. Ferrari, A., Cachia, R. and Punie, Y. (2009). *Innovation and Creativity in Education and Training in the EU Member States: Fostering Creative Learning and Supporting Innovative Teaching*. JRC Technical Note, 52374.
41. Reid, A. and Petocz, P. (2004). Learning domains and the process of creativity. *The Australian Educational Researcher*, 31(2): 45–62.
42. Nonaka, I. (1991). The knowledge-creating company. *Harvard Business Review*, 69(6): 96–104.
43. Santonen, T., Kaivo-oja J. and Suomala, J. (2007). *Introduction to National Open Innovation System (NOIS) Paradigm. A Preliminary Concept for Interchange*, FFRC eBooks 8/2007. Finland Futures Research Centre, Turku School of Economics, Turku, 37.
44. Shirouzu, H. Miyake, N. and Masukawa, H. (2002). Cognitively active externalisation for situated reflection. *Cognitive Science*, 26: 469–501.
45. See West *et al.* (2014), pp. 805–806.
46. Murphy, S. A. and Kumer, V. (1997). The front end of new product development: A Canadian survey. *R&D Management*, 27: 5–15.
47. Jetter, A. (2003). *Educating the Guess: Strategies, Concepts and Tools for the Fuzzy Front End of Product Development*. IEEE, pp. 261–273.
48. Osborn, A. F. (1963). *Applied Imagination*, 3rd revised edition. New York, NY, US: Charles Scribner's Sons.
49. Furnham, A. (2000). The brainstorming myth. *Business Strategy Review*, 11(4): 21–28.

50. Rietzschel, E. F. (2005). *From Quantity to Quality: Cognitive, Motivational and Social Aspects of Creative Idea Generation and Selection.* PhD Thesis. Utrecht, Utrecht University.
51. Toubia, O. and Florès, L. (2007). Adaptive idea screening using consumers. *Marketing Science*, 26(3): 342–360.
52. Girotra, K., Terwiesch, C. and Ulrich, K. T. (2010). Idea generation and the quality of the best idea. *Management Science*, 56(4): 591–605.
53. Hammedi, W. A., Van Riel, C. and Sasovova, Z. (2011). Antecedents and consequences of reflexivity in new product idea screening. *Journal of Product Innovation Management*, 28(5): 662–679.
54. Stroebe, W., Diehl, M. and Abakoumkin, G. (1992). The illusion of group effectivity. *Personality and Social Psychology Bulletin*, 18: 643–650.
55. Paulus, P. B., Kohn, N. W. and Arditti, L. E. (2011). Effects of quantity and quality instructions on brainstorming. *The Journal of Creative Behavior*, 45(1): 38–46.
56. Mullen, B., Johnson, C. and Salas, E. (1991). Productivity loss in brainstorming groups: A meta-analytic integration. *Basic and Applied Social Psychology*, 12(1): 3–23.
57. Diehl, M. and Stroebe, W. (1991). Productivity loss in idea-generating groups: Tracking down the blocking effect. *Journal of Personality and Social Psychology*, 61(3): 392.
58. Paulus, P. B., Dzindolet, M. T., Poletes, G. and Camacho, L. M. (1993). Perception of performance in group brainstorming: The illusion of group productivity. *Personality and Social Psychology Bulletin*, 19: 78–89.
59. Shalley, C. E., Zhou, J. and Oldham, G. R. (2004). The effects of personal and contextual characteristics on creativity: Where should we go from here? *Journal of Management*, 30(6): 933–958.
60. Hogle, J. G. (1996). *Considering Games as Cognitive Tools: In Search of Effective Edutainment.* ERIC Clearinghouse.
61. Bredemeier, M. E. and Greenblat, C. S. (1981). The educational effectiveness of simulation games: A synthesis of findings. *Simulation & Games*, 12(3): 307–332.
62. Dempsey, J., Lucassen, B. Gilley, W. and Rasmussen, K. (1993). Since Malone's theory of intrinsically motivating instruction: What's the score in the gaming literature? *Journal of Educational Technology Systems*, 22(2): 173–183.
63. Gardner, H. (1983). *Frames of Mind* (2nd ed.). (New York: Basic Books).
64. Jacobs, J. W. and Dempsey, J. V. (1993). Simulation and gaming: Fidelity, feedback, and motivation. In Dempsey, J. V. and Sales, G. C. (eds.), *Interactive Instruction and Feedback*, Englewood Hills, NJ: Educational Technology Publications, pp. 197–227.
65. Seginer, R. (1980). Game ability and academic ability: Dependence on S.E.S. and psychological mediators. *Simulation and Games*, 11(4): 403–421.

66. Fontana, L. A. Dede, C. White, C. S. and Cates, W. M. (1993). Multimedia: A gateway to higher-order thinking skills. *Paper Presented at the Convention of the Association for Educational Communications and Technology*, New Orleans.
67. Smith, K. J. (1992). Using multimedia with Navajo children: An effort to alleviate problems of cultural learning style, background of experience, and motivation. *Reading & Writing Quarterly: Overcoming Learning Difficulties*, 8(3): 287–294.
68. Turner, S. V. and Dipinto, V. M. (1992). Students as hypermedia authors: Themes emerging from a quantitative study. *Journal of Research on Computing in Education*, 25(2): 187–199.
69. Wilson, K. S. (1991). Two multimedia design research projects: Palenque and the museum visitor's project. *Paper Presented at the Society for Applied Learning Technologies* (Washington, DC).
70. Deci, E. L. and Ryan, R. M. (1985). Intrinsic motivation and self-determination in human behavior (NewYork, Plenum).
71. Lepper, M. R. and Chabay, R. (1985). Intrinsic motivation and instruction: Conflicting views on the role of motivational processes in computer-based education. *Educational Psychologist*, 20(4): 217–230.
72. Middleton, J. A. and Toluk, Z. (1999). First steps in the development of an adaptive theory of motivation. *Educational Psychologist*, 34(2): 99.
73. Utman, C. H. (1997). Performance effects of motivational state: A meta-analysis. *Personality and Social Psychology Review*, 1: 170–182.
74. Covington, M. and Müeller, K. (2001). Intrinsic versus extrinsic motivation: An approach/avoidance reformulation. *Educational Psychology Review*, 13(2): 157–176.
75. Malone, T. and Lepper, M. (1987). Making learning fun: A taxonomy of intrinsic motivations for learning. In Snow, R. and Farr, M. (eds.) *Aptitude, Learning, and Instruction*, Vol. 3. Hillsdale, NJ: Earlbaum, pp. 223–253.
76. Amabile, T. M. (1998). How to kill creativity. *Harvard Business Review*, 76(5): 76–87.
77. Ryan, R. M. and Deci, E. L. (2000). Intrinsic and extrinsic motivations: Classic definitions and new directions. *Contemporary Educational Psychology*, 25(1): 54–67.
78. Ryan, R. M. (1982). Control and information in the intrapersonal sphere: An extension of cognitive evaluation theory. *Journal of Personality and Social Psychology*, 43: 450–461.
79. Harter (1981). A new self-report scale of intrinsic versus extrinsic orientation in the classroom. *Developmental Psychology*, 17(3): 300–312.
80. Etzkowitz, H. and Leydesdorff, L. (2000). The dynamics of innovation: From national systems and 'Mode 2' to a Triple Helix of university–industry–government relations. *Research Policy*, 29: 109–123.

81. Etzkowitz, H. and Leydesdorff, L. (1999). The future location of research and technology transfer. *Journal of Technology Transfer*, 24(2–3): 111–123.
82. Gergils, H. (2005). *Dynamic Innovation Systems in the Nordic Countries? A Summary Analysis and Assessment* (Stockholm, SNS förlag, 2005).
83. Ranga, M. and Etzkowitz, H. (2012). Great expectations: An innovation solution to the contemporary economic crisis. *European Planning Studies*, 20(9): 1429 –1438.
84. Ranga, M. and Etzkowitz, H. (2013). Triple Helix systems: Analytical framework for innovation policy and practice in the Knowledge Society. *Industry and Higher Education*, 27(4): 237–262.
85. Hughes, B. B. and Hillebrand, E. E. (2006). Exploring and Shaping International Futures. Ifs Setup V7, January 2014 (Paradigm Publishers, Boulder, Co. Pardee Center, University Denver, Denver, Co.).
86. Pardee Center, IFs Setup V7, January 2014 (University Denver, Denver, Co., 2006).
87. Takeuchi, H. and Nonaka, I. (1986). The new product development game. *Harvard Business Review*, Jan–Feb: 137–146.
88. Kaivo-oja, J. (2006). *Towards Integration of Innovation Systems and Foresight Research in Firms and Corporations. The Classical Takeuchi–Nonaka Model Reconsider and Reformulated* (Finland Futures Research Center. FFRC Publications 2/2006. Turku School of Economics, Turku).
89. Leiponen, A. and Drejer, I. (2005). Technical regimes and strategy: Intra-industry heterogenity in the organization of innovation activities. A Paper Presented in the DRUID Tenth Anniversary Summer Conference 2005 on *Dynamics of Industry and Innovation Organizations, Networks and Systems.* Theme B: Technical Change and Corporate Dynamics (DRUID, Denmark).
90. Rohrbeck, R. and Gemünden, H. G. (2011). Corporate foresight: Its three roles in enhancing the innovation capacity of a firm. *Technological Forecasting and Social Change*, 78(2): 231–243.
91. Johnston, B. D. (2011). *Science Fiction Prototyping. Designing the Future with Science Fiction, Synthesis Lectures on Computer Science #3* (Morgan and Claypool Publishers).
92. Carrott, J. H. and Johnson, B. D. (2013). *Vintage Tomorrows. A Historian and A Futurist Journey Through Steampunk into The Future of Technology* (Sebastopol, CA).
93. Huiban, J. P. and Boushina, Z. (1988). Innovation and the quality of labour factor. *Small Business Economics*, 10: 389–400.
94. Inauen, M. and Schenker-Wicki, A. (2012). Fostering radical innovations with open innovation. *European Journal of Innovation Management*, 15(2): 212–231.
95. Ståhle, P. A. Pöyhonen and Kyläkeiko, K. (2004). Towards operationalization of dynamic capabilities. In Hosni, Y. (ed.), *Internet Economy: Opportunities*

*and Challenges for the Developed and Developing Regions of the World* (Elsevier, Amsterdam).

96. Haho, P. (2002). Strategic and Operational Process Innovations in NPD Processes Applying Evolutionary Change Management, (4th International CINet 2002 Conference, "*Continuous Innovation in Business Processes and Networks*", Espoo, September 1518, 299–314.
97. Urabe, K. (1988). Innovation and the Japanese Management System. In Urabe, K., Child, J. and Kagono, T. (Eds.), *Innovation and Management, International Comparisons* (De Gruyter and Co, Berlin), pp. 3–25.
98. Dewar, R. D. and Dutton, J. E. (1986). The adoption of radical and incremental innovations: An empirical analysis. *Management Science*, 32(11): 1422–1433.
99. Henderson, R. M. and Cla, K. B.
100. Yu, D. and Hang, C. C. (2010). A reflective review of disruptive innovation theory. *International Journal of Management Reviews*, 12(4): 435–452.
101. Abernathy, W. J. and Clark, K. B. (1985). Innovation: Mapping the winds of creative destruction. *Research Policy*, 14(1): 3–22.
102. Lynn, G. J., Morone and A. Paulson (1996). Marketing and discontinuous innovation: The probe and learn process, *California Management Review*, 38(3), Spring 1996. Available at SSRN: http://ssrn.com/abstract=2151914.
103. Ali, A. (1994). Pioneering versus incremental innovation: Review and research propositions. *Journal of Product Innovation Management*, 11(1): 46–61.
104. Mascitelli, R. (2000). From experience: Harnessing tacit knowledge to achieve breakthrough innovation. *Journal of Product Innovation Management*, 17(3): 179–193.
105. Danneels, E. (2004). Disruptive technology reconsidered: A critique and research agenda. *Journal of Product Innovation Management*, 21(4): 246–258.
106. Christensen, C. M. (1997). *The Innovator's Dilemma: When New Technologies Cause Great Firms to Fail* (Harvard Business School Press, Boston, Mass.).
107. Cho, H. and Pucik, V. (2005). Relationship between innovativeness, quality, growth, profitability and market value. *Strategic Management Journal*, 26(5): 555–575.
108. Nelson, R. R. and Winter, S. G. (1982). *An Evolutionary Theory of Economic Change* (Belknap, Cambridge, MA).
109. Teece, D. J., Pisano, G. and A. Shuen (1997). Dynamic capabilities and strategic management. *Strategic Management Journal*, 18(7): 509–533.
110. Leonard-Barton, D. (1992). Core capabilities and core rigidities: A paradox in managing new product development. *Strategic Management Journal*, 13(8): 111–125.
111. O'Connor, G. Ravichandran, T. and Robeson, D. (2008). Risk management through learning: Management practices for radical innovation success. *Journal of High Technology Management Research*, 19: 70–82.

112. Oerlemans, L. A. G., Knoben, J. and Pretorius, M. W. (2013). Alliance portfolio diversity, radical and incremental innovation: The moderating role of technology management. *Technovation*, 33: 234–246.
113. Büschgens, T., Bausch, A. and Balkin, D. B. (2013). Organizing for radical innovation — A multi-level behavioral approach. *Journal of High Technology Management Research*, 24: 138–152.
114. Colarelli Connor, G. and McDermon, C. M. (2004). The human side of radical innovation. *Journal of Engineering and Technology Management*, 21: 11–30.
115. Lettl, C. (2007). User involvement competence for radical innovation. *Journal of Engineering and Technological Management*, 24(1–2): 53–75.
116. Sainio, L.-M., Ritala, P. and Hurmelinna-Laukkanen, P. (2012). Constituents of radical innovation — Exploring the role of strategic orientations and market uncertainty. *Technovation*, 32: 591–599.

# Chapter 7

# LEVERAGING DESIGN THINKING TO INNOVATE

Denis Dennehy, Frederic Adam and Fergal Carton

*Lero — The Irish Software Research Centre*
*JE Cairnes School of Business & Economics*
*National University of Ireland Galway (NUIG)*
*Galway, Ireland*
*denis.dennehy@nuigalway.ie*

Advances in mobile technology and the shift from product-based economies to service-based economies are forcing companies to create new value propositions which inevitably impact the design of their business model. In response, companies are seeking new approaches to adapt to this ever changing complex environment. Design thinking is increasingly being viewed as a new approach to dealing with complex challenges in business, most notably when applied to areas such as product and service innovation, strategy development and ultimately to achieving competitive advantage. This chapter explains what design thinking is and explores how it complements the practice of management by providing empirical evidence on the benefits that practitioners have from a wide range of industries in Ireland and Luxembourg experienced when they adopted a positive design attitude when using visualisation tools to approach a complex business challenge.

*Keywords*: Design Thinking; Business Model; Innovation.

## 1. Introduction

With increasing growth in the service-based economy and advances in mobile technologies, companies are being challenged to *"fundamentally rethink their business models and radically transform their capabilities"* (p. 7).[1] In contrast to the 20th century, when value creation was achieved by organisations implementing linear improvements (e.g., supply chain management) which were usually defined by the dialog of heuristics (understanding of a mystery) to algorithms (a formula that could be driven to huge scale and scope), value creation in the 21st century will be defined more by the conversion of mysteries to heuristics.[1] Figure 1 illustrates three activities that organisations need to move through to find a successful formula (product/service).

The Knowledge Funnel illustrates how knowledge moves through the funnel. The first stage is the exploration of a *mystery* which can take on many forms (how would a new mobile payment system look). The second stage of the funnel is *heuristic*, a particular way that enables an organisation to narrow the scope of inquiry, to make the mystery a manageable size and a simplified way of understanding

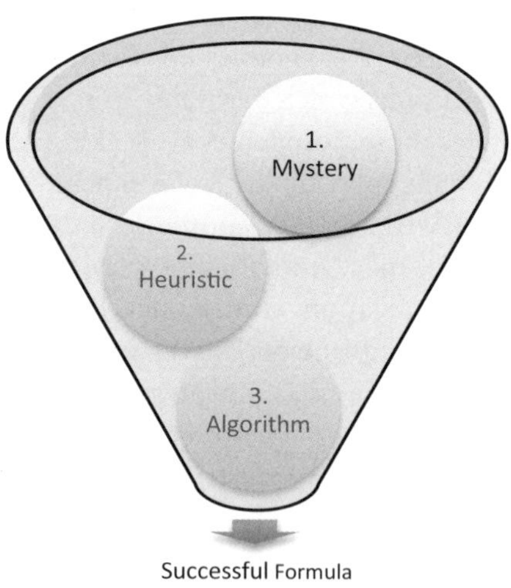


**Figure 1.** The Knowledge Funnel

*Source*: Adapted from Ref. [2].

**Table 1.** Characteristics of exploration and exploitation

| | Exploration | Exploitation |
|---|---|---|
| Organisational focus | The invention of business | The administration of business |
| Overriding goal | Dynamically moving from the current knowledge stage to the next | Systematically honing and refining within the current knowledge stage |
| Driving forces | Intuition, feeling, hypothesis about the future, originality | Analysis, reasoning, data from the past, mastery |
| Future orientation | Long-term | Short-term |
| Progress | Uneven, scattered, characterised by false starts and significant leaps forward | Accomplishment by measured, careful incremental steps |
| Risk and reward | High risk, uncertain but potentially high reward | Minimal risk, predictable but smaller rewards |
| Challenge | Failure to consolidate and exploit returns | Exhaustion and obsolescence |

*Source*: Ref. [2], p. 20.

that will enable people to focus their efforts, and the third stage is the *algorithm* stage. Algorithm is when the company puts its heuristic into action, studies it, thinks about it in detail and converts into a successful formula (moving a pilot mobile payment system to a national/universal mobile payment system). However, finding and repeating this successful formula is difficult for companies as it requires two very different activities to move through the knowledge funnel which involves *'exploration'* for new knowledge and *'exploitation'* of existing knowledge.[3] The characteristics of these two activities are listed in Table 1.

Although the use of design thinking and visualisation tools has become increasingly popular in recent years, their use is not new[4] but have much to offer management. However, Leavy[3] (p. 6) expresses a concern that *"the design of business introduces a set of concepts that are not familiar to most managers"*. Further, to become a design-thinking company is not easy because most companies have a built-in bias to rely on analytical thinking and exploitation, as well as favour reliability over validity.[2,3] The optimal approach to reliability and validity is the same as exploration and exploitation, balancing both rather than choosing one over the other which can be achieved by applying design thinking. Figure 2 illustrates how design thinking can enable companies to balance their analytical thinking (reliability) and their intuitive thinking (validity).

Most managers make decisions using deductive reasoning (the logic of what must be) and inductive reasoning (the logic of what is operative) which are both linked with analytical thinking and are appropriate to exploiting existing knowledge.[3] However, a third form of reasoning known as *abductive reasoning* (what might be) is more appropriate for the exploration of new knowledge. It is through abductive reasoning that organisations can leverage both analytical thinking and intuitive thinking.[2] Design-thinking companies do not only rely on deductive and inductive reasoning to solve business problems, they also use abductive reasoning. Having identified how conventional management thinking primarily relies on analytical thinking to address business problems, the remainder of this chapter is structured as follows. First, we distinguish between two different types of business problems (Section 2). Following this

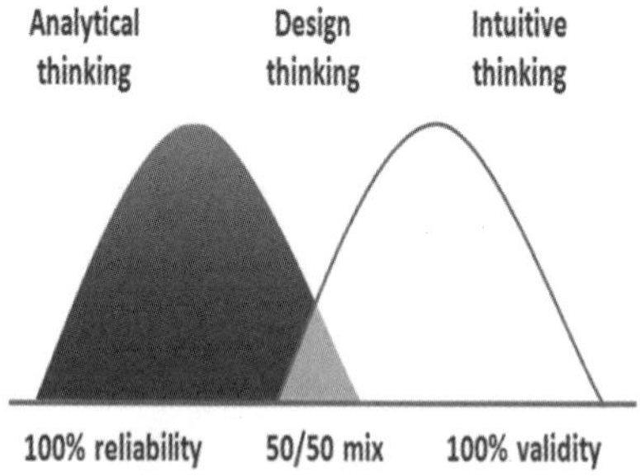


**Figure 2.** The predilection gap

*Source*: Ref. [2].

is a discussion on how design thinking and the use of visualisation tools can complement the practice of management (Section 3). Then, justification for adopting the research methodology is provided (Section 4). Section 5 presents the key findings of our research. A conclusion is provided in Section 6 followed by the implications to the practice of management (Section 7). Limitations of the study are acknowledged in Section 8. The final section, Section 9 provides recommendations for action in the future.

## 2. Distinguishing Between Wicked and Tame Problems

Conklin and Weil[5] advocate that it is important to distinguish between tame (simpler) problems and wicked problems because tame problems require the traditional linear process of problem-solving methods (understand the problem which can include data gathering and analysis, then formulate and implement a solution) which are not appropriate to solving wicked problems. Buchanan[6] (p. 15) defines wicked problems as *"a class of social system problems which are ill-formulated, where the information is confusing, where there are many clients and decision-makers with conflicting values, and where the ramifications of the whole system are confusing"*. As it is difficult to recognise a wicked problem, as many problems appear tame[5], it is important to identify the characteristics of each of these two types of problems. The characteristics of tame and wicked problems are listed in Table 2.

Adding to the complexity of a wicked problem in the context of business (for example, a project) is that most wicked problems involve lots of stakeholders[5] which are listed in Table 3.

It is because of these stakeholders that Conklin and Weil[5] (p. 7) believe that solving a wicked problem is essentially a social process and that *"a linear approach to solving a wicked problem simply won't work"*. Nowadays, because most problems in business are wicked or ill-structured[7], Conklin[8] (p. 3) states that *"failing to recognise the 'wicked dynamics' in problems, we persist in applying*

**Table 2.** Characteristics of tame and wicked problems

| | Tame problems | Wicked problems |
|---|---|---|
| 1 | Has a well-defined and stable problem statement. | Unstable requirements and constraints. |
| 2 | Has a definite stopping point (i.e., when the solution is reached). | Complex interactions among subcomponents of the problem and resulting subcomponents of solution. |
| 3 | Has a solution which can be objectively evaluated as right or wrong. | Inherent flexibility to change artifacts and processes. |
| 4 | Belongs to a class of similar problems which are all solved in the same similar way. | Dependence on human cognitive abilities (e.g., creativity). |
| 5 | Have solutions which can be easily tried and abandoned. | Dependence on human social abilities (e.g., teamwork). |

**Table 3.** Possible stakeholders in a wicked problem

| | Type of stakeholders |
|---|---|
| 1 | All members of the project team. |
| 2 | Upper management. |
| 3 | People in other parts of the organisation working on related projects. |
| 4 | People in other departments, like finance or purchasing, who have some general oversight function. |
| 5 | External stakeholders, such as customers, investors, partner companies, regulators, watchdog organisations, and organisations in other countries. |

*Source*: Ref. [5].

*inappropriate methods and tools to them*". More worrying is that Dunne and Martin[9] (p. 518) posit that conventional management thinking "*views constraints as an undesirable barrier to the generation and implementation of ideas*" and that managers avoid wicked problems because their source of status comes from elsewhere in the organisation. Having identified the differences between tame and wicked problems, Section 3 explains how design thinking complements the practice of management.

## 3. How Design Thinking Complements the Practice of Management

Design thinking is increasingly being viewed as a new approach to deal with complex challenges in business, most notably when applied to business management, strategy development and sustaining a unique competitive advantage.[10–13] A similar supporting view of design thinking is expressed by Leavy[3] (p. 6) who advocates that *"design thinking, or the creative principles long associated with the design function, may now have something very significant to offer when applied more broadly to business management and strategy development"*. Design thinking can make an influential difference in human-centered activities such as communication and collaboration, as well as business model innovation.[12,14] Design thinking assumes the human experience is always messy, and sees true objectivity as an illusion, and answers are either better or worse.[15] Oster[16] (p. 109) defines design thinking as *"a process for developing numerous practical alternate solutions to specific problems or issues fully using the individual and group capabilities and ideas of employees, customers, vendors, and the general public"*. The differences between business thinking and design thinking are presented in Table 4.

Table 4 acknowledges the subjective nature of the human experience and emphasises the importance of iteration to find a better answer and to move between abstract and particular levels of focus. The notion of design thinking in the practice of management is not new as Boland *et al.* (p. 11)[17] remind us that the Nobel Laureate Herbert Simon *"declared the centrality of designing to managers"* and refer to his book, *The Sciences of the Artificial*, as *"one of the finest examples of what we call design attitude for managers"* because Simon *"equated managing with decision-making and argued*

**Table 4.** Differences in business and design thinking

| | Business thinly | Design thinly |
|---|---|---|
| Underlying assumptions | Rationality, objectivity; reality as fixed and quantifiable | Subjective experience; reality as socially constructed |
| Method | Analysis aimed at proving one 'best' answer | Experimentation aimed at iterating toward a 'better' answer |
| Process | Planning | Doing |
| Decision drivers | Logic; numeric models | Emotional insight; experiential models |
| Values | Pursuit of control and stability; discomfort with uncertainty | Pursuit of novelty; dislike of *status quo* |
| Levels of focus | Abstract or particular | Iterative movement between abstract and particular |

*Source*: Ref. [15].

*that there are three essential aspects of decision-making: intelligence, design and action*"[17] (p. 12). More recently, Avital and Te'eni[18] (p. 154) extends this view of management by stating that "*management is not just an act of decision-making between a given set of alternatives; it is also the active, ongoing shaping and designing of organisations and their stakeholders' experiences.*" These views of management reassert the value of design thinking within the practice of management. This reassertion is supported by Martin[1] who identifies three important implications for this shift in focus on how companies create value in the 21st century and how managers approach and solve wicked problems: (1) Design skills and business skills are converging, (2) the need for new kinds of business enterprise, and (3) a change of focus in our thinking about design and business. In order to adapt to this shift in how companies do business, companies will need to shift from being a *traditional firm* to a *design shop*. The comparisons between a traditional firm and a design shop are listed in Table 5.

**Table 5.** Comparison of traditional firms to design shops

| Feature | From "Traditional firm ..." | ... to "Design Shop" |
|---|---|---|
| Flow of work life | Ongoing tasks<br>Permanent assignments | Projects<br>Defined terms |
| Source of status | Managing big budgets and large staffs | Solving 'wicked problems' |
| Style of work | Defined roles<br>Wait until it is 'right' | Collaborative<br>Iterative |
| Mode of thinking | Deductive<br>Inductive | Deductive<br>Inductive<br>Abductive |
| Dominant attitude | We can only do what we have budget to do<br>Constraints are the enemy | Nothing cannot be done<br>Constraints increase the challenge and excitement |

*Source*: Ref. [1], p. 9.

In order to shift from a traditional firm to a design shop, business people *"don't need to understand designers better: they need to be designers. They need to think and work like designers, have attitudes like designers, and learn to evaluate each other as designers do"*[1] (p. 10). Design thinking can be viewed as a discipline that *"uses a designer's sensibility and methods to match the people's needs with what is technologically feasible and what a viable business strategy can convert into customer value and market opportunity"* (p. 85).[14]

Avital *et al.*[4] (p. 154) advocate that the benefits of design thinking in the practice of management is that it *"encourages a constructive, divergent behaviour that guards managers against premature closure in decisions and actions"* because this approach is *"action oriented and brings together people from different disciplines and practices to jointly pursue a vision"*. A similar supporting view is expressed by Raney and Jacoby[13] (pp. 35–36) who advocate that *"when tackling the most abstract and ambiguous challenges, design thinking can lead to more effective decisions"* as well as *"help reduce risk and convert decision-making meetings from consensus building slogs to collaborative, invigorating critique and build sessions"*. While the espoused benefits of design thinking are compelling, its use has not become mainstream in the practice of management. One possible reason why design thinking has not become mainstream in the practice of management, except for the engineering discipline which is familiar with design techniques, is that business oriented stakeholders are usually unaware of such activities.[19] Yet, many organisations use visualisation tools (business model canvas, service blue printing) to develop new products and services and to achieve a competitive advantage. Using visualisation to address business problems is discussed in Section 3.

### 3.1. *Using visualisation to address business problems*

Representing information externally can prevent or reduce an overload of mental resources and it can "*contribute to a more detailed problem analysis*" because the information is stored externally[20] (p. 145). The visual representation of information "*offers many cognitive (*e.g., *overview), emotional (*e.g., *inspiration) and social (*e.g., *common focus) advantages that can be put to use in the business strategy process*" (p. 4).[21] Nickerson *et al.*[22] (p. 2) also believe that "design, diagrams, sketches, visualisations, and the like play an essential role in many cognitive activities". A similar supporting view is expressed by Moody[23] (p. 765) who advocates that "*visual representations are deemed to be effective because they tap into the capabilities of the powerful and highly parallel human visual system*". Tversky[24] (p. 500) believes that the power of leveraging visuals in all forms is that they externalise our thoughts for various reasons and often simultaneously:

- — Some are aesthetic: to arouse emotions or evoke pleasure;
- — Some are behavioural: to affect action or promote collaboration;
- — Some are cognitive: to serve as reminders, to focus our thoughts; to reorganise our thoughts, and to explore thoughts;
- — And many are communicative: to inform both self and others.

Visuals that are developed within the course of a meeting assist participants to focus on the current state of the conversation, while also acting as a memory aid of what has already being discussed and what are the outstanding issues.[25,26] In addition, visuals continuously remind participants of the "big picture to which they contribute"[26] (p. 89) and essentially become a fundamental tool supporting conversations, while also assisting the stakeholders to develop mutual understanding.[25]

Shostack[27] (p. 139) states that unlike verbal definitions, visualisation "encourages creativity and pre-emptive problem solving II ... and enhances management's ability to think effectively". A similar supporting view is shared by Jonassen[7] who believes that the key to problem-solving is problem representation. In contrast to verbal representations, visual representations (i.e., service blueprints, business modelling) are more accurate, are less inclined to be misinterpreted and can reduce the potential for failure as it encourages creativity, pre-emptive problem solving and controlled implementation.[28] For example, although service blueprinting shares similar characteristics (i.e., using visual notations to represent actors and activities) with other process modelling techniques (i.e., Unified Modelling Language, Business Process Modelling Notation), it is less formal and complex which enables all stakeholders to meet particular requirements.[28] Though visualisation has been an essential tool in the analysis of social networks, visualisations of complex systems such as service value networks in a complete and readable manner is challenging.[29] Section 4 discusses the research approach that was adopted in this study.

## 4. Research Methodology

In this study we adopted a design science research approach to create the visualisation' tools and leveraged the focus group technique to evaluate the tools. Although there are different views on what constitutes a design science study, there is consensus that any design science study must produce and evaluate a novel artifact.[30–33] As the process of design science research requires the frequent iteration between the development and evaluation of the artifact, Tremblay *et al.*[34] recommend a pilot focus group, followed by at least two exploratory focus groups (EFGs) and two confirmatory focus groups (CFGs). Between January and August 2014, we conducted a pilot focus group, three exploratory focus groups and two confirmatory focus groups with over 80 practitioners from diverse industry backgrounds in Ireland and Luxembourg. Using focus groups was appropriate as this technique is frequently used to diagnos the potential for problems with a new program, service, or product, when generating impressions of products, services, programs, institutions, or other objects of interests, and when learning how respondents talk about the phenomenon of interest.[35] Focus groups also provide a number of benefits which include: Flexibility to handle a wide range of topics, the researcher has direct interaction with respondents, large amounts of rich qualitative data, building on other respondent's comments, multiple perspectives, and are more naturalistic than face-to-face interviews.[34–37] Although there are many compelling reasons why focus groups are an appropriate evaluation technique in design science, there are, however, some limitations with the technique. These limitations include: The small numbers of participants can limit generalisation to a larger population, and the moderator may knowingly or unknowingly provide cues that could bias the results.[35]

We followed the recommended two hour duration for each focus group[35] during which time participants were presented with a descriptive scenario.[38] The scenario (business problem) was how to position multiple stakeholders involved in the design of a mobile payment business model. Participants were required to approach this scenario (business problem) first by relying on their dominant analytical thinking and then by using a visualisation tool in the form of a canvas. The hypothesis was that in order to utilise such a tool, participants need to apply the characteristics of a design thinker. In doing so, participants were able to balance their analytical thinking and intuitive thinking. A synthesis of our findings from across the six focus groups is presented in Section 5.

## 5. Key Findings

The use of design thinking and visualisation tools in business has sometimes been associated with people from design backgrounds. An objective of the study was to identify what benefits (if any) do practitioners realise when they adopt a design thinking to approach the complex challenge of managing partnerships in an m-payment scenario. In doing so, the findings presented in this section demonstrate that managers from non-design backgrounds share the innate characteristics and attitude of a designer. A design attitude anticipates that each project is a new opportunity and that the conditions (e.g., beliefs, expectations, practices, and policies) of a new project situation must be respected.[17] At the heart of a designer's attitude is the "*willingness to embrace discontinuity as well as the open-ended character of the design process, in which ambiguity prevails and outcomes are uncertain*"[39] (p. 241). Avital *et al.*[18] (p. 157) believe that managers can "*emulate designer's work practices, and engage in design discourse with new vocabularies that reflect a design attitude*". Boland and Collopy[40] (p. 13) define a design attitude as "*a thorough, ongoing expectation that each project is a new opportunity to create something remarkable, and to do it in a way that has never been done before*".

Table 6 provides evidence of the benefits that practitioners experienced when using design thinking and visualisation tool (Partnership Management Canvas) to approach a complex business problem. Table 6 adds credence to the adage '*a picture is worth a thousand words*' as it provides managers with a holistic view of the problem to be addressed and it provides the potential to create a shared understanding amongst the stakeholders involved in finding a suitable solution. In addition, realising the benefits listed in Table 6 is important because "*during face-to-face interactions, groups frequently overly rely on the dominant viewpoint to lead the group in its decision-making process*" (p. 1).[41]

These finding are valuable because Bitner *et al.*[28] posit that because services are intangible, people frequently use words only to specify the requirements. As a result, stakeholders rely on the use of

Table 6. How design thinking supports practitioners to approach complex challenge

| Promoted visibility | Provided a common platform | Promoted interaction | Encouraged a shared commitment | Promoted discussion | Encouraged problem-solving | Created a shared understanding |
|---|---|---|---|---|---|---|
| *"It made everything more visible and raised issues that may arise"* | *"Common language and focus good for all stakeholders' understanding of the process"* | *"A vehicle for brainstorming"* | *"Ensured the correct level of involvement at the right time and good to have a model to bring together all parties"* | *"Everyone had a voice"* | *"Helped to identify customer value proposition and to focus all stakeholders in the group"* | *"Identification of gaps and could be used to exclude some stakeholders"* |
| *"Ability to see each partner's contributions and motivations"* | *"Ability to listen to others, ability to contribute objectively, ability to negotiate"* | *"Good interactions amongst partners"* | *"Collaboration amongst team was better"* | *"Good discussion tool and looking at specifics of problem"* | *"Helps focus on issues"* | *"Very good grounding on the project for all partners"* |
| *"Better visualisation of roles and responsibilities to be shared with new partners joining the ecosystem"* | *"Delivering a message to the other stakeholders"* | *"Able to agree more"* | *"Main one was team-building"* | *"Discussion, clarification, communication"* | *"The visual ability to be able to work as a team collaborating, challenging each other's skills"* | *"Helped identify roles and responsibilities of partners and how they change over time"* |

**Table 7.** Benefits of design thinking and visualisation tools

| | Benefits gained |
|---|---|
| 1 | Provided a neutral platform as a basis for structured discussion that enabled all partners to have a voice. |
| 2 | Promoted visibility for each partner's contribution, their motivations and their shared responsibilities. |
| 3 | Identified issues and knowledge gaps that had not surfaced previously. |
| 4 | Encouraged pre-emptive problem-solving at an early stage. |
| 5 | Improved understanding of complexities of the business challenge. |
| 6 | Encouraged collaboration and attainment of mutual agreement amongst partners. |
| 7 | Stimulated interaction and communication between the partners. |
| 8 | Identified the changing roles and responsibilities of stakeholders. |

natural language requirement representations which has a number of shortcomings[42] such as the inclusion of irrelevant information, omission of valuable information, over-specification, vagueness, contradictions and wishful thinking.[43]

A synthesis of the benefits experienced by practitioners during the course of this study is presented in Table 7.

Having provided compelling evidence on the benefits of using design thinking and visualisation tools in the practice of management. Section 6 provides a conclusion to the study.

## 6. Conclusions

The study first established that although the practice of management is paralleled with decision-making which consists of three indispensable features, intelligence, design and action[33], most managers are unaware of decision by design and continue to rely on the dominant form of logical decision-making. To address this lack of awareness, the study advocated that managers could apply design thinking to their practice by adopting a positive design attitude which would enable them to develop a holistic perspective to approach a wicked problem by opening communication and interaction with other stakeholders, and include all voices of stakeholder in an interactive fashion.

Leveraging design thinking and visualisation tools to approach a wicked problem was appropriate as design thinking accepts that the human experience is always messy.[15] In addition, a review of design thinking literature identified a number of benefits that can be gained from design thinking and visualisation tools in the practice of management. These benefits include the encouragement of creativity, pre-emptive problem-solving and enabling people from diverse business backgrounds to come together and collectively pursue a shared vision. This study demonstrated that when practitioners apply a positive design attitude to approach a wicked problem, they experience a number of benefits: Increased visibility of the problem, a common platform for communication and discussion, increased interaction between stakeholders and greater focus on problem-solving. Realising these benefits is highly beneficial as organisations are increasingly learning to collaborate with multiple partners from different industries to achieve a sustainable competitive advantage. Key to realising and aligning the benefits of all stakeholders inevitably depends on the stakeholders having a neutral platform that will enable them to collectively pre-empt problems and solutions and to create a shared understanding that will lead to a shared commitment. Following this conclusion, Section 7 highlights the implications to the practice of management.

## 7. Managerial Implications

By providing a conducive environment to engage with over 80 experienced practitioners from diverse backgrounds, this study demonstrated that managers have an innate capacity to leverage design thinking and use it to complement and balance their analytical thinking (e.g., reliability) and their intuitive thinking (e.g., validity). This study provided empirical evidence on the benefits that were realised by practitioners when they used the Partnership Management Canvas and applied design thinking to approach a complex business challenge. By realising these benefits, this study challenges the myopic criticism that has been directed towards the business model concept and associated visualisation tool (e.g., business model canvas) as being static in a dynamic business environment. Rather than view such tools as static, this study reasserts that when organisations shift from a traditional thinking firm to a design firm, they will tap into a manager's natural design thinking ability to effectively coordinate the resources and competencies of the organisation, as well as having a positive impact on the strategic positioning of the organisation. Further, these benefits show that when practitioners are encouraged to use design thinking to approach complex business challenges, then the efficacy of such tools are realised and a win–win outcome can be achieved by all stakeholders. This is a valuable contribution to the practice of management as it takes the mystery out of design thinking and places it at the core of how organisations in the 21st century can operate. The second contribution to the practice of management is established in Section 8. As with any research projects, there are limitations of this study and are acknowledged in Section 8.

## 8. Limitations of the Study

As with any research project, there are limitations. This research project is no exception as there are two limitations. The *first* limitation is that although the visualisation tool was used to address a wicked problem in the context of this study, we cannot claim that this specific tool can address similar wicked problems in other instances because how organisations solve wicked problems is influenced by the organisations' culture and design competence of managers. The *second* limitation is that the researcher adhered to the recommended two-hour duration for conducting a focus group; there was limited time to gain a deeper understanding about the mobile payment phenomenon. Future research could extend the duration of the focus group technique to suit the requirements of the organisation responsible for solving the business problem. We provide recommendations for future work in the next and final section of this chapter.

## 9. Recommendation for Action in the Future

This section provides two important recommendations for future work that is aimed at academia and practice. *First,* as the study highlighted that practitioners have an appetite for such visualisation tools in a digital format, future research could focus on developing software that enhances collaboration and understanding between multiple stakeholders. *Second,* future work for organisations and managers would be able to view the application of design thinking and the use of visualisation tools as equal importance when compared to other forms of work practices because advances in technology and the evolution of human behaviours will continue to present new wicked problems more frequently. It is how organisations approach and manage these wicked problems that will inevitably determine the innovators and market leaders in m-payment value networks in the 21st century.

## References

1. Martin, D. (2004). The design of business. *Rothman Management*, 7–11.
2. Brown, T. (2009). *The Design of Business* (Harvard Business Press, Boston, Massachusetts, USA).
3. Leavy, B. (2010). Design thinking — A new mental model of value innovation. *Strategy & Leadership*, 38(3): 5–14.
4. Crapo, A. W., Waisel, L. B., Wallace, W. A. and Willemain, T. R. (2000). Visualization and the process of modeling: A cognitive-theoretic view. *In Proceedings of the Sixth ACM SIGKDD International Conference on Knowledge Discovery and Data Mining,* pp. 218–226.
5. Conklin, J. and Weil, W. (2007). Wicked problems: Naming the pain in organizations. *White Papers,* pp. 1–13. http://www.accelinnova.com/docs/wicked-problems.pdf (accessed 21-10-14).
6. Buchanan, R. (1992). Wicked problems in design thinking. *Design Issues* 8(2): 5–21.
7. Jonassen, D. (2003). Using cognitive tools to represent problems. *Journal of Research on Technology in Education*, 35(3): 362–381.

8. Conklin, J. (2001). *Wicked Problems and Social Complexity. Dialogue Mapping: Building Shared Understanding of Wicked Problems.* http://www.ideapartnership.org/documents/wickedproblems.pdf (accessed 19-10-14).
9. Dunne, D. and Martin, R. (2006). Design thinking and how it will change management education: An interview and discussion. *Academy of Management Learning & Education*, 5(4): 512–523.
10. Clark, K. and Smith, R. (2008). Unleashing the power of design thinking. *Design Management Review*, 19(3): 8–15.
11. Dorst, K. (2011). The core of 'design thinking' and its application. *Design Studies*, 32(6): 521–532.
12. Fraser, H. (2006). Turning design thinking into design doing. *Rotman Magazine* (Spring/Summer): 24–28.
13. Raney, C. and Jacoby, R. (2010). Decisions by design: Stop deciding, start designing. *Rotman Magazine*, 34–39.
14. Brown, T. (2008) Design thinking. *Harvard Business Review* 86(6): 84–96.
15. Liedtka, J. and Ogilvie, T. (2011). *Designing for Growth — A Design Thinking Toolkit for Managers* (Columbia University Press, New York, USA).
16. Oster, G. (2008). Practitioners corner: Derailing design thinking. *International Journal of Leadership Studies*, 4(1): 107–115.
17. Boland, R. (2002). Managing as Designing: Creating a Vocabulary for Management Education and Research. *Frontiers of Management Workshop*, Weatherhead School of Management, Cape Western Reserve University, USA, June 14–15, pp. 106–112.
18. Avital, M. and Te'eni, D. (2009). From generative fit to generative capacity: Exploring an emerging dimension of information systems design and task performance. *Information Systems Journal*, 19(4): 345–367.
19. Gordijn, J. and Akkermans, J. (2003). Value-based requirements engineering: Exploring innovative e-commerce ideas. *Requirements Engineering*, 8(2): 114–134.
20. Römer, A., Leinert, S. and Sachse, P. (2000). External support of problem analysis in design problem solving. *Research in Engineering Design*, 12(3): 144–151.
21. Eppler, M. J. and Platts, K. W. (2009). Visual strategizing: The systematic use of visualization in the strategic-planning process. *Long Range Planning*, 42(1): 42–74.
22. Nickerson, J. V., Corter, J. E., Tversky, B., Zahner, D. and Rho, Y. J. (2008). The spatial nature of thought: Understanding systems design through diagrams. *In Proceedings of Intenational Confernece on Information Systems* (ICIS) 2008. Paper 216.
23. Moody, D. (2009). The "physics" of notations: Toward a scientific basis for constructing visual notations in software engineering. *Software Engineering, IEEE Transactions*, 35(6): 756–779.
24. Tversky, B. (2011). Visualizing thought. *Topics in Cognitive Science*, 3(3): 499–535.

25. Kraut, R. E., Fussell, S. R. and Siegel, J. (2003). Visual information as a conversational resource in collaborative physical tasks. *Human–Computer Interaction*, 18(1): 13–49.
26. Mengis, J. and Eppler, M. J. (2006). Seeing versus arguing the moderating role of collaborative visualization in team knowledge integration. *Journal of Universal Knowledge Management*, 1(3): 151–162.
27. Shostack, G. L. (1984). Designing services that deliver. *Harvard Business Review* 62: 133–139. http://www.semanticfoundry.com/docs/servicesThatDeliver.pdf (Last accessed 18/10/2014).
28. Bitner, M. J., Ostrom, A. L. and Morgan, F. N. (2008). Service blueprinting: A practical technique for service innovation. *California Management Review*, 50(3): 66–94.
29. Brandes, U., Raab, J. and Wagner, D. (2001). Exploratory network visualization: Simultaneous display of actor status and connections. *Journal of Social Structure*, 2(4): 1–28.
30. Hevner, A., March, S. T., Park, J. and Ram, S. (2004). Design science in Information Systems Research. *MIS Quarterly*, 28(1): 75–105.
31. March, S. T. and Smith, G. F. (1995). Design and natural science research on information technology. *Decision Support Systems*, 15(4): 251–266.
32. Peffers, K., Tuunanen, T., Rothenberger, M. A. and Chatterjee, S. (2007). A design science research methodology for information systems research. *Journal of Management Information Systems*, 24(3): 45–77.
33. Simon, H. A. (1969). *The Sciences of the Artificial* (MIT Press, Cambrigde, *Massachusetts*, USA).
34. Tremblay, M. C., Hevner, A. R. and Berndt, D. J. (2010). Focus groups for artifact refinement and evaluation in design research. *Communications of the Association for Information Systems*, 26: 599–618.
35. Stewart, D. W., Rook, D. W. and Shamdasani, P. N. (2007). *Focus Groups: Theory and Practice*. Sage Publications, London, UK.
36. Wilkinson, S. (2004). *Qualitative Research: Theory, Method and Practice*. Sage Publications, London, UK.
37. Peffers, K. and Tuunanen, T. (2005). Planning for IS applications: A practical, information theoretical method and case study in mobile financial services. *Information & Management*, 42(3): 483–501.
38. Peffers, K., Tuunanen, T., Gengler, C. E., Rossi, M., Hui, W., Virtanen, V. and Bragge, J. (2006). The design science research process: A model for producing and presenting information systems research, pp. 83–106. *In Proceedings of the First International Conference on Design Science Research in Information Systems and Technology (DESRIST 2006)*, California, USA.
39. Michlewski, K. (2008). Uncovering design attitude: Inside the culture of designers. *Organization Studies*, 29(3): 373–392.

40. Boland, R. J. and Collopy, F. (2004). *Design matters for management.* In Boland, R. J. and Collopy, F. (eds.), Managing as Designing (Stanford: Stanford University Press).
41. Dimicco, J. M., Pandolfo, A. and Bender, W. (2004). Influencing group participation with a shared display. In *Proceedings of the 2004 ACM Conference on Computer Supported Cooperative Work*, pp. 614–623.
42. Finkelstein, A., Kramer, J., Nuseibeh, B., Finkelstein, L. and Goedicke, M. (1992). Viewpoints: A framework for integrating multiple perspectives in system development. International Journal of Software Engineering and Knowledge Engineering, 2(01): 31–58.
43. Meyer, B. (1985). On formalism in specifications. *IEEE Software*, 2(1): 6–26.

# Chapter 8

# OPEN INNOVATION ADOPTION PRACTICES AND EVALUATION METHODS IN THE GLOBAL PROCESS INDUSTRY

Jarkko Pellikka,* Miika Kajanus[†] and Marko Seppänen[‡]

**Nokia Technologies, Labs*
*P.O. Box 226, FI-00045 Nokia Group*
*Karaportti 4, 02610 Espoo, Finland*
*jarkko.pellikka@nokia.com*
*[†]Savonia University of Applied Sciences, Kuopio*
*miika.kajanus@savonia.fi*
*[‡]Tampere University of Technology, Department of Pori*
*P.O. Box 300, FI-28101 Pori, Finland*
*marko.seppanen@tut.fi*

Inter-organisational collaboration is crucial for Open Innovation adoption in order to realise, in practice, the potential benefits of the concept. However, the number of empirical studies on the activities and processes that contribute to Open Innovation adoption is relatively limited. Therefore, this chapter provides empirical evidence from the global process industry, where we have defined, tested and utilised the standardised process, evaluation methodologies and tools that engage several collaborators to improve the key business process performance of their firms. The results show that the standardised process can provide several concrete business results for the management, and can efficiently promote the adoption of Open Innovation, as well as the utilisation of the firm's tacit knowledge to further improve the business performance.

*Keywords*: Open Innovation; Adoption; Multiple-Criteria Decision Support; Process Industry.

## 1. Introduction

The capability to create new innovations is important for both large and small companies in order to sharpen competitiveness and enhance growth.[1] However, a firm must have an organisational structure and processes that encourage the effective generation and commercialisation of innovation.[2] According to Schilling[3], a firm operating in a high-technology field needs: (1) In-depth understanding of innovation dynamics; (2) a well-crafted innovation strategy; and (3) well-designed processes for innovation with the external collaborators in order to bring in complementary assets.

It is important to note that since innovation has become one key source of competitive advantage, companies have also become increasingly dependent on external sources of knowledge (e.g., technological collaboration) rather than rely only on in-house research and development activities. Challenging competitive situations and shorter product lifecycles force companies in the different industries to generate innovations more quickly and reduce their cost structure in research and development, while seeking external complementary resources from the public and private sectors as a part of their Open Innovation strategy execution. Such inter-organisational collaboration processes can provide a strong basis for the generation and commercialisation of different types of innovation, as well as other potential benefits, such as developing manufacturing and other business operations to new markets.[4,5]

Previously, managers, researchers and policymakers have made increasing efforts to identify ways to foster and support business growth and development in modern, dynamic business environments in order to promote growth.[6] In addition, rather than relying on internal resources for processes, companies are increasingly utilising an 'Open Innovation' approach,[7] which is defined in this chapter as an innovation-related, inter-organisational collaboration between business partners, which contributes to companies' innovation process. This, however, means that the companies should be able to create partnerships with external technology providers and other partner companies (thus building a network of inter-organisational R&D including several actors, such as customers, business partners, universities, etc.) who can all provide resources and knowledge assets[8] to utilise Open Innovation utilisation.

Previous studies have also pointed out the importance of better understanding multi-level (firm, industry and business environment) Open Innovation practices.[9] For example, one relatively unknown area in this context is the practical process (workflow) between the counterparts in order to realise the benefits of the Open Innovation approach in the industrial context, which seems to be an essential part of the wider business process management and business development framework.[10] In addition, we currently have only some preliminary results on the Open Innovation approach's practical usage and the role of executives in the practical implementation activities.[11] Therefore, this chapter empirically demonstrates how the Open Innovation approach can be implemented in practice in the global process industry by using a standardised process that engaged several external business partners (private and public organisations) to seek, evaluate and decide on further business development activities. Thus this chapter has two objectives:

- To describe how Open Innovation can be efficiently implemented and utilised in the global process industry via standardised and facilitated processes.
- To provide a documented workflow and concrete methods to realise the benefits in the global process industry via the Open Innovation approach.

Therefore, the main contribution of this chapter is to provide the empirical evidences on the practical adoption of the Open Innovation approach within the global process industry. The structure of the chapter is as follows: In Section 2, we highlight the importance of further elaborating concrete ways to utilise Open Innovation via Business Process Management perspective. In Section 3, we present the foundation for the created Open Innovation adoption process. Section 4 provides information on the 6 case studies that were conducted during the research to test the process and here we also discuss about the key results and the realised benefits from the case companies. Concluding remarks and suggestions for future studies are presented in Section 5.

## 2. Open Innovation Adoption and Business Process Management

### 2.1. *Importance of knowledge-based resources*

The foundation of the Open Innovation adoption is that it potentially integrates a number of experts, collaborators and customers into the innovation processes.[12,13] According to Huizingh[14], the inter-organisational collaboration may have different types of structures and parties, including internal and external collaborators and their cross-functional engagement to reach the targets set for the new innovation development. Especially, the role of knowledge and knowledge-based resources has become an essential part of these inter-organisational structures.[15–17] In particular, the success of firms in increasingly turbulent environments depends heavily on the quality of knowledge-based assets and their successful application to develop business performance via business processes.[18] According to Teece[19], these activities may help companies support the following initiatives:

(1) Introduction of the new innovative products that may further help companies increase productivity and generate business growth.[20]
(2) Mobilisation of employees' skills and motivation for continuous improvement in process capabilities, quality and response times.

(3) Further development of customer relationships.[21] It is still important to note that knowledge has no value if it is not applied in some way; therefore, companies should acquire knowledge from external sources to bring together diverse inputs and specialised knowledge.[22]

It is also important to note that knowledge, ideas or valuable resources do not directly ensure profits unless they are effectively implemented.[23] However, this requires a definition of the process that can both capture the need to select the 'best' ideas for implementation in a context in which several internal cross-functional teams, departments and/or units, as well as inter-organisational collaborators, are involved to improve business performance. Although there are some practical examples and best practices on the Open Innovation adoption processes[24,25], our understanding of the adoption activities, tools and methods in the inter-organisational context is still relatively limited. Adopting the Open Innovation concept within an inter-organisational context requires, for example, concrete practices and processes (including roles and responsibilities), to ensure successful cross-functional adoption within the organisation.[26] Therefore, we applied Business Process Management (BPM) as a key foundation in our study to reach the research objectives. The utilised conceptual foundation of BPM is described in Section 2.2.

## 2.2. *The concept of Business Process Management (BPM)*

The dynamic business environment has forced firms to continuously seek to improve their business processes. The target is to enhance business performance via efficient business processes[21,27] and BPM.[28] The term 'process' is typically used in the academic literature[29] to signify, for example, an organisation's activities that convert the specific inputs to outputs, via certain workflow and decision-making practices, to reach the strategic business objectives. In addition, according to Davenport and Short[30], business processes typically have two important characteristics: (1) Customers for the business process output can be either internal or external to the organisation; (2) cross-organisational and cross-functional boundaries are in place, which means that business processes are generally independent of formal organisational structure, and typically involve several organisations and internal business functions. Taking the above definitions into account, in this chapter, we use the term 'business process' as a pre-defined procedure that is required to deliver value (defined business output) to an organisation and customers, via the assigned resources, in order to achieve both the organisation's short- and long-term business objectives.[30,31]

In this chapter, the term 'Business Process Management' (BPM) is defined to include an organisation's efforts to continuously and cross-functionally manage and improve business processes to realise business benefits related to BPM utilisation. For example, Aguilar-Savèn[32] pointed out that business process models are mainly used to: (1) Learn about the business process; (2) make decisions on the business process; and/or (3) develop business process software. The following business benefits have also been found:[33–35]

(i) Business process descriptions and documentation that help to manage and improve productivity (e.g., via business process optimisation).
(ii) Identification and analysis of the improvement opportunities that are based, for example, on the process measurements and process scorecards.
(iii) Visualisation of the roles, responsibilities and cross-functional and inter-organisational linkages to create a transparent structure for the business process planning, management and control.
(iv) Both qualitative and quantitative decision-making models and tools for continuous process improvement.

Based on the approaches presented above, we expect that the BPM perspective will provide a promising foundation for companies to adopt Open Innovation in order to provide a structure and governance for the collaborators. In the next section, we will present the operational approach that has been created and piloted with the key research partners in the pilot project. The use of pilots has been previously highlighted to be very useful in an Open Innovation context as it increases our understanding of how things actually work (or do not work) via innovation-related business processes among the inter-organisational collaborators.[14] The use of the cases here is particularly essential since Open Innovation is currently more frequently adopted by high-technology companies in contexts that may be characterised by globalisation, technology intensity, technology fusion, new business models and knowledge leveraging, rather than by low- and medium-technology companies.[17,36] This research project was conducted in a process industry that is typically understood to be a medium-technology industry rather than a high-technology industry. However, the process industry is still very global by nature, and in some areas very sophisticated technology plays a crucial role (e.g., in manufacturing processes). Therefore, we also assume that the following sections will contribute to the current Open Innovation literature from a new angle.

## 3. Data and Methods

### 3.1. *Overview of the methods*

Multiple-Criteria Decision Support (MCDS) methods are a family of methods that provide transparent and coherent decision support to facilitate communication in complex situations characterised by conflicting objectives, multiple alternatives and actors, and uncertain or insufficient information. As in this chapter, MCDS methods have been integrated with other planning tools. For example, connecting MCDS methods to SWOT (Strengths, Weaknesses, Opportunities and Threats) analysis yields analytical priorities for the SWOT factors, which makes them commensurable.[37] The purpose of using MCDS methods within an Open Innovation framework is to systematically assign importance to each identified alternative. First, through a creative process (e.g., brainstorming), a set of potential alternatives were identified.

As highlighted in the literature review above, concepts and information from a large range of disciplines, such as innovation management, business process management and organisation theory, need to be taken into account when examining the adoption of Open Innovation at the operational level. As already described, companies are dependent on inter-organisational collaboration, external knowledge-based resources and, especially, the capability to efficiently utilise these assets in Open Innovation activities. Since there was no universal model for the utilisation of Open Innovation, a qualitative approach based on the case study method was adopted in our study.

### 3.2. Case studies, Pain Chain Analysis and Lean Six Sigma

Previous authors have reported that case studies can potentially help to analyse and improve business performance and innovation activities.[38,39] Furthermore, firm-level case studies can increase understanding of the innovation process and help build more robust analyses of innovation utilisation at the operational level.[40] Therefore, we utilised the concept of Pain Chain Analysis as a basic framework within which an analysis of the decision situation in the case studies was performed (see detailed descriptions of the cases in Table 1). All the case studies were conducted during 2012 and 2013 in Finland, France, and the United States.

In general, Pain Chain Analysis utilises the Lean Six Sigma approach, which enables organisations to effectively utilise methods and tools to define, measure, analyse, improve and control business processes to meet customers' needs.[41,42] In addition, the Lean Six Sigma approach can be effectively utilised in order to identify and remove causes of defects, improve lead time and productivity and reduce operational costs.[43]

The justification for Lean Six Sigma utilisation as a basis for the Pain Chain Analysis concept was based on the following arguments. Lean Six Sigma helps organisations to improve customer requirements, business systems, productivity, and financial performance[44],

**Table 1.** Case study descriptions, identified main challenges, results, benefits, and company commentaries

| | CASE 1 | CASE 2 | CASE 3 | CASE 4 | CASE 5 | CASE 6 |
|---|---|---|---|---|---|---|
| Industry context and location | Paper industry, Finland | Paper industry, France | Bioeconomy cluster, Finland | Regional development company, Finland | Wood product company, USA | Electricity and telecommunication constructor, Finland |
| General description | Product uniform development in a whole production chain via staff group theme interviews and analytical evaluation process. | Paper mill benefits from increased energy efficiency over 12% via pilot case. | Optimising program evaluation via visual and analytical evaluation process and tools. | A joint development plan for the wood processing firms in the region. | Business model design for European markets via analytical business model design tool. | Open innovation for developing cost efficiency in value chain of telecommunication infrastructure construction. |
| Background description | The mill is one of the leading integrated paper mills. Therefore, the main target of the pilot project was to improve the product uniform by decreasing volatility in the whole production chain. The key objectives of the pilot project were: (1) Define the key areas in the quality management and product uniform for further development actions to increase the productivity and business performance; | The mill is one of the leading mills in environmental management. Therefore, the main target of the pilot project was to improve the current measurement system on energy efficiency even further in two paper machines. The key objectives of the pilot project were: (1) Define the key areas in energy efficiency for further development actions to increase productivity; | Cluster's research program was part of the 5-year resource efficiency research theme started in June 2008. The goal was to improve the competitiveness of the whole value chain. The Cluster program focused on developing radically new energy- and resource-efficient web production technologies, and reengineering the product concept of fibre-based products. The research program consists of 10 projects. | The company is a regional development company facilitating regional business development. Wood product industry is one of the important industrial areas in the region. | The company is a medium-sized US manufacturer of value-added wood products that is desired to design a business model for export market entry. The objective for the exercise was to prepare a Vision for New Business Development: Strategic Planning for Future Export Sales. | Company is one of the leading constructors of utility projects for electricity and telecommunications in Finland. The main target was to improve the productivity of the whole value chain of Telco infrastructure construction by using co-creation method with the customer. The key objectives were: Define the discontinuity points and savings from the end-to-end delivery process for the further development actions to increase productivity and business performance. |

*(Continued)*

**Table 1.** *(Continued)*

| | CASE 1 | CASE 2 | CASE 3 | CASE 4 | CASE 5 | CASE 6 |
|---|---|---|---|---|---|---|
| | (2) Prioritise the most promising development areas in improving product uniform; and (3) Identify the feasible solutions and actions to improve both customer and personnel satisfaction. | (2) Prioritise the most promising development areas in energy savings and energy efficiency; and (3) Identify the cost effective solutions and/or mitigation actions for further development in energy efficiency. | | | | Prioritisation of the most promising development areas for improving productivity. To identify the feasible solutions and actions to improve both customer and personnel satisfaction. |
| Main challenge | To facilitate own staff and involved partners to identify feasible and significant improvements in an efficient and straightforward way. | To control and manage unexpected variances in the energy efficiency measurements. In addition, it was assumed that the current measurement system is not fully optimised to provide accurate information for the site management. | To utilise transparent and systematic approach on program evaluation by creating the evaluation approach that reaches all the pre-set goals for the evaluation process (i.e., highly analytical and descriptive, standardised and easy to use for the senior executives). | To identify and agree what are the main activities to be implemented jointly by wood industry companies and other regional actors. | To clarify export strategies since the company has been doing business in domestic markets only and has had only single actions in export markets. | To find common development items which make strong link to interest and scope of development strategy of actors in the complex business environment. |
| The key results and process steps to reach solution | (1) 11 group interviews (43 persons); (2) 102 development ideas, which concerned areas such as information flows, different way of actions, education and investment needs; (3) Ideas evaluated by 17 experts against selected criteria. | (1) 26 individual development ideas identified during the analysis, based on the site interviews; (2) Robust Portfolio Modelling and comprehensive data analysis were used to identify TOP5 idea; | (1) ***Project ratings***: Evaluators rated the importance of each project against the criteria; (2) ***Scenarios***: A number of scenarios were composed by using scoring model with different combinations criteria weights; | (1) Definition of the decision context and making the preparations for a planning process such as selecting and briefing the participants; (2) Three development themes were identified; | (1) Definition of the decision context and making the preparations for a planning process; (2) 5 executives from the company and 4 outside experts participated; | (1) Kick-off meeting and phone interviews organised, totally 40 persons participated in interviews and generated 93 development ideas; (2) 25 project ideas were evaluated by 18 experts against the selected criteria. |

| | | | | | |
|---|---|---|---|---|---|
| Robust Portfolio Modelling was employed to identify the best ideas. TOP22 ideas were identified; (4) Key persons participated in a half-day workshop; (5) Three action plans (with concrete goals, schedule and responsibilities) based on the TOP22 ideas. | (3) Improvement plan, which indicates key activities to further improve the total energy efficiency, was developed. | (3) ***Analysis***: The projects were analysed by calculating core Indices with different constraint parameters; (4) Final results visualise and simulate the impact of the changes and the critical variables on actions in strengthening the future projects. | (3) 17 companies were interviewed and 88 ideas for the three development themes were identified; (4) Ideas were elaborated in three theme group workshops and the final set of the ideas was evaluated by the key companies; (5) Robust Portfolio Modelling was employed to identify the ideas that best meet the criteria; (6) The set of the ideas was selected, and the action plan with concrete goals, schedule and responsibilities elaborated. | (3) The workshop resulted in the list of business model items (90 items altogether) for which the company executives determined the relative importance; (4) Most important business model items were analysed by core index calculation; (5) The research team analysed the evaluation results and produced two different, consistent options for business models; (6) These were discussed and further developed together with company people. | (3) Robust portfolio modelling were employed to identify ideas which best fulfilled the needs of different criteria; (4) TOP11 ideas were identified; (5) Key persons from the company and customer participated in a half-day workshop; (6) Three action plans with concrete goals, schedule and responsibilities based on the TOP11 ideas were the outcome. |

(*Continued*)

**Table 1.** *(Continued)*

| | CASE 1 | CASE 2 | CASE 3 | CASE 4 | CASE 5 | CASE 6 |
|---|---|---|---|---|---|---|
| Benefits perceived during and after the pilot project | The results of the pilot show improvements in productivity and the perceived competitive advantage. The identified benefits are based on the better and faster information flows and action in the whole production chain. In addition, more optimised daily production via modern automation solutions helps the site management to maintain and further develop the production and product uniform in the future. | The pilot project provided several concrete benefits for the program's management group: (1) Visual, analytical and systematic evaluation process; (2) Transparent and efficient way to identify the best projects in the project portfolio; (3) Easy to use with the multiple evaluators. | | The joint action plan for regional wood product industry was the concrete outcome. The plan has development actions covering areas such as increasing the use of timber in wood products, increasing the know-how and competitive advantage of the industry, and customer-oriented and international business development. | The process resulted in better business model understanding and faster creation of the business model; it was finalised as an action plan. The tool helped the executives gain the results faster and with less perceived mental effort. | The concrete development items were clearly considered, including approximate calculation of saving potential to all projects of action plan. In addition, the company and its direct customer started a systematic co-creation forum for develop productivity. |
| | The pilot project also enabled 43 cross-functional team members from the site to share their broad experiences and remarkable know-how on the further production process development. The forum of these professionals is also playing a key role in order to fully realise the benefits in the focus areas. | (4) Enable the utilisation of all the decision-making criteria and all the possible decision-making criteria combinations (weights) at the same time (core values). | | The action plan provides also a tool to facilitate regional development actions. The pilot project also enabled 17 companies to share their broad experiences and remarkable know-how on the further business development of wood product industry. | Participative approach in the method helps participants to gain results by providing an opportunity for effective and fast cooperation with business model experts. Two alternative business models were designed, and for both models an action | The company also receives valuable information and know-how of co-creation tools and methods. |

| | | | | | | |
|---|---|---|---|---|---|---|
| | | | (5) It indicates the critical variables for which additional information may be obtained. The executives can directly consider actions that help to optimise the future projects. | This forum is playing a key role in realising the benefits in the focus areas. | plan for future company activities was generated. The pilot project also enabled five executives to share their broad experiences and remarkable know-how on the further business model development for new market entry. | |
| Company comments | *'The process helped us to identify a good set of promising ideas and to select the best of them to implement in advance. The implemented actions improved our competitive advantage and the process itself improved information flows among production chain and staff's motivation for continous improvement'. — Mill Manager* | *'The comprehensive understanding of the complexity and the factors that impact on our energy efficiency is very critical. After the pilot project we know the well-defined items for the further energy efficient improvement'. — Mill Technical Director* | *'The interface of the evaluation tool was very descriptive and visual. It was essential that we managed to evaluate all these good projects analytically and via systematic way to find the best of the best'. — Member of the Program Management Group* | *'The process helped us to collect the development needs and the ideas for the future actions from the companies and other key actors in the region. The process resulted in the joint action plan'. — Regional Cluster Business Manager* | *'The process helped us to create the business model vision and to finalise the action plan. Process increased our understanding of business model thinking and helped us to gain the results faster and with less mental effort'. — Company CEO* | *'The tool and process provided a systematic and efficient open innovation framework for co-creation in value chain of telecommunication infrastructure construction. Process helped us to identify potential discontinuity points and savings from the multi-actor process. Evaluation with strong math helps participants to make priority project list for action plan'. — Development and Quality Manager* |

which were the key business objectives among the pilot companies in the process industry. In addition, Lean Six Sigma was especially needed to seek new opportunities for further product quality improvement, keeping in mind the currently very high-level quality of the pilot companies. In each pilot project conducted in the research project, the effective implementation was assumed to require an open mindset about cultural changes in the pilot organisations, from the shop floor to the senior management, as well as new and innovative approaches to optimise production via inter-organisational and cross-functional collaboration, and a high degree of systematic, hands-on participation.[45]

In practice, Pain Chain Analysis was used to further exploit the Lean Six Sigma approach. Pain Chain Analysis included four main steps, in order to gather ideas to improve the manufacturing process performance:

(1) Preliminary planning phase in which the key objectives of the Pain Chain Analysis were defined together with the site senior management.
(2) Focus Group interview planning that included the planning activities, such as semi-structured theme group interview planning (content) and scheduling of the interviews.
(3) Conduction of the focus group interviews at the sites.
(4) Detailed analysis of the interview data and preparation of the data for the evaluation (e.g., identification of the categories and specific ideas).

The output of the Pain Chain Analysis was the finalised evaluation in which all the process development ideas were listed and described in detail, and made ready for the evaluators in the evaluation environment. The main phases of Pain Chain Analysis can be described in Figure 1.

This output initiated the next phase, where the identified ideas were then prioritised by evaluating them analytically. Any MCDS method, including its prioritisation principles, can be applied, and the existence of different techniques allows the adaptation of use of

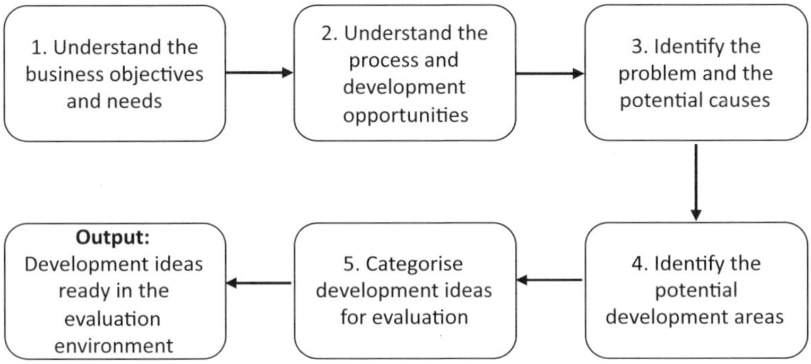


**Figure 1.** Process of Pain Chain Analysis

the MCDS method according to the needs of the decision-maker and the specific planning situation. The Open Innovation process typically closely parallels project portfolio selection[46–48], in which the organisation seeks to choose alternatives (projects) that meet its objectives and resource constraints. Thanks to extensive methodological research, there exists a broad variety of approaches to project portfolio selection. For reviews of project portfolio selection methods, see Gustafsson and Salo[49] and Martino.[50]

Robust Portfolio Modelling (RPM)[51,52] is an MCDS method designed to select ideal sets of projects (or items) to constitute a portfolio. In its basic variant, RPM is a scoring model[53] which is widely employed in the evaluation and selection of projects (portfolio of projects) in settings where multiple objectives are required.[54] A central concept and key outcome of RPM is Core Indices. Core Index values are used to map information and identify ideal portfolios of items. Core index values can support Open Innovation processes by indicating alternatives that are most capable of fulfilling the contrasting needs.

## 3.3. *Evaluation and Robust Portfolio Modelling*

The overall process of Open Innovation, facilitated by the use of MCDS tools, generally occurs in four primary phases:

(i) Definition of the decision context, and making the preparations for a planning process, such as selecting and briefing the participants.
(ii) The relevant alternatives are identified and included in the analysis. Pain Chain analysis can be applied.
(iii) The relative importance of each alternative is determined using an MCDS method.
(iv) The selection of the alternatives to be implemented with the aid of the evaluation information.

Phase 1 is needed to define the context for the business situation in order to make the process effective. This includes, for example, deciding on the objective for the process, identifying participants (for example, deciding whether potential customers or other collaborators are needed), and identifying criteria to be used in the evaluation, etc. Phase 2 can be completed using many different approaches. Typically, it is done through some form of group ideation where the executives of the firm (and possibly other participants) collectively consider what is important with respect to the situation. As in all creative processes, it is important to differentiate the creative search for possible items from their analysis. The final result of this phase is a list of alternatives that might be important in the business situation.

In Phase 3, the list of alternatives developed in Phase 2 is prioritised. An MCDS method is employed in this prioritisation process using the following steps. Phase 2 may result in many items (e.g., >75). Each evaluator rates the importance of each Phase 2 item against criteria selected by the project team. For example, the evaluator may be asked to rate the importance of the item as 'providing competitive advantage' or 'offering business potential'.

Using an MCDS method is one way of making sense of the many ratings generated in Phase 2. Any MCDS method, including its prioritisation principles, can be applied according to the needs of the specific planning situation. In this case, we performed a method resulting from the calculation of so-called Core Indices. The core index indicates in how many effective portfolios this particular item is included. A value of 1 indicates that the item is present in all portfolios whereas a value of 0 indicates it is none. If the item falls between 1 and 0, it is said to be borderline'. A narrow versus wide constraint parameter, which identifies the number of items to be included in the portfolio, can be employed. In the case of using a narrow constraint parameter, few items are included in effective portfolios; widening the parameter includes more items. In Figure 1, items 1 and 2 are included in both of the optimal portfolios in the case of a narrow constraint parameter (in the middle) whereas in the case of wide constraint parameter, items 1, 2, 7, 14, 16, and 26 are included in both of the optimal portfolios (the right hand side).

Finally, Phase 4 is where the actual action plan is formulated based on the information produced in previous phases. This is where the practical judgment of the research/consulting team, in concert with firm executives, must be exercised. A qualitative process is used to create one or more potential action plans. More information about the evaluation tool and the evaluation phase can be found via the following link: https://www.youtube.com/watch?v=xHaEeSmHmS8.

## 4. Results and Discussion

In this research, we implemented the described process in six different case studies. The conducted processes are described in Table 1. As Table 1 shows, the created and conducted process for Open Innovation adoption can concretely enable several collaborators to contribute to business performance development and innovation. The methodological approaches, tools and activities have provided a basis for inter-organisational knowledge-sharing among the collaborators in order to create different types of new innovations in the process industry at the firm level (Cases 1, 2, 5, and 6), at the regional level (Case 4) and at the cluster level (Case 3). In general, the starting points in each case study were to lean, streamline and optimise current business processes in the individual organisation (see Cases 1, 2, and 5) and/or between several organisations (see Cases 3, 4, and 6). Therefore, the standardised approach, based on BPM and Lean Six Sigma approaches, was very beneficial to efficiently respond to the business requirements (see the problem statement in Table 1).

All of the case studies started with a clear definition of the problem statement that was based on the business requirements in each case. The main purpose of the clear problem statement was to describe a general pain in the current state and then transform it into a well-defined problem with clarity and focus. In most of the cases, the problems highlighted the difference between the current and desired outcome (i.e., future state) and also acted as a starting point for the following steps (e.g., interview planning), bringing alignment and purpose to inter-organisational cross-functional improvement team and the key stakeholders.

The clear problem statement required strong support from the top-management of the each organisation. As pointed out in several previous studies[55], top leaders of the organisation play a key role by creating consistent conditions and allocating resources for the creation and implementation of new innovation to support business performance. In the beginning of all cases, we arranged several sessions with the top management in order to define the scope of work and also to make sure that the management understood their role as a change agent and motivator throughout the process.

After the conducted pilot projects, feedback was collected from the firm executives. The feedback indicated that the tool and the process employed worked in this case, resulting in better business understanding, faster creation and the finalised action plan for implementation. The executives emphasised that the tool and the process helped them gain the results faster and with less mental effort. All in all, the experience encourages further development of the business model design tool. It can be concluded that one of the benefits of this methodology is its participative approach. The method helps participants to gain results and effectively utilise cross-functional knowledge to develop, for example, manufacturing process (see Cases 1 and 2). However, the experience and the feedback indicated several issues which could be improved.

The importance of the Phase 1 decision context and preparation of the process was emphasised. The decision context should be worked out carefully before the first workshop. In this case, a remarkable share of the time spent in Workshop 1 went toward reaching joint understanding about the firm context. Also, questions from participants are important to consider; for example, if key customers or key partners should be involved in the process. Participating experts asked questions in order to clarify understanding and to bring new points of view into the discussion. Utilising outside experts in this way can be seen as a step towards Open Innovation practice. It can also accelerate the whole business model creation process.

The evaluation of numerous items towards several criteria was considered time-consuming and laborious by the executives. This was in spite of the fact that a user-friendly visual tool was used. The evaluation of the items performed a key element in the process, allowing the core index analysis. Three of the five executives executed the evaluation. There are two ways to make this easier. First, minimise the number of items to be evaluated by putting more emphasis on the selection of the items in the previous workshop. Second, develop further the evaluation process and tool. During this study, the evaluation was done separately on two criteria. One idea could be to employ two-dimensional evaluations at the same time. The visual evaluation tool used in this case supports the use of smaller evaluation categories; for example, the items to be evaluated could be classified into several evaluation categories, which would make the evaluation process easier for the evaluator. The evaluation tool supports adding written comments, but no such comments were received from the evaluators. The reason for this is obviously the short evaluation time and large number of items to be evaluated.

Here we used approximated core index values instead of employing calculations based on the RPM-methodology.[51] This affects one source of errors in our process. It might happen that, especially in the case of low CP, the calculation method we used does not identify all the items included for an optimal portfolio. For example, an item might be included in an optimal portfolio, if criteria weights are 78.6%/21.4% and not 70%/30% or 74%/25%. However, when the constraint parameter is increased, those items are included. Moreover, designing the final business model (see Case 5) is based on the expertise of the planning team, and all necessary items are included whether they are identified by the calculation method or not. That is why we considered that our approximated core index values were sufficient to illustrate the methodology. However, if this kind of business model design process is performed, it is recommend to use core index calculations based on RPM-methodology or similar, which mathematically assures identification of all potential core items.

Parts of this development process could be performed in one day. We feel that if properly executed, a one-day workshop could result in outputs similar to those described above. This would be possible with the effective use of participatory development methods and tools, as well as with careful planning of the participatory creation process. Also, full commitment from all stakeholders would be essential.

## 5. Conclusion and Implications

The main objective of this chapter was to demonstrate with concrete results how Open Innovation adoption can be performed via inter-organisational collaboration in the global process industry. The findings contribute to prior studies by providing multi-level results on standardised structure for Open Innovation adoption within a context in which several external collaborators are needed in order to develop the current state. As pointed out in this chapter, acquiring knowledge and information among the collaborators is essential in innovation activities to improve the outcome of the process.[56]

From the managerial perspective, the findings mean that the predefined and documented process for Open Innovation adoption, including the key activities, milestones and deliverables, enables top management to gather, analyse and exploit the inter-organisational, knowledge-based resources to create new innovations, develop leaner manufacturing and improve business performance.[57] In order to reach these targets, management can utilise a BPM approach to build the management system, key processes, tools and control measures for innovation management purposes with the collaborators.[58,59]

As previous studies have clearly highlighted, the management of the organisation plays a key role in enabling the innovation-related activities between the collaborators to take place.[60] However, this requires that the organisation needs to first decide what are the key capabilities and knowledge-based resources to create innovation, how much of these capabilities exist internally and what are the key areas where the organisation needs external capabilities and complementary assets.[61] After that, the partner selection process can be started with the candidates before the Open Innovation adoption process can be started, as described in this study. The managers should take this into account, especially from the schedule perspective. In addition, the partner selection within the adoption process (e.g., external interviewees and evaluators) should be carefully considered in order to find motivated and committed individuals with the relevant capabilities to engage the process.

There are some attractive opportunities for future research. For example, among future studies examining the Open Innovation adoption in a multi-level context, it would be valuable to follow the progress after the creation of the implementation plan in order to see the long-term results. As noted earlier, the realisation of the value of innovation may take several years to see the real impact. Therefore, a longitudinal study may provide very beneficial information about the real value of the process described in this paper, and would also enable greater understanding of on the complex linkage between innovation and business performance.

## Acknowledgements

We would like to thank Tuomo Eskelinen from Savonia University of Applied Sciences and Mikko Heinonen from Voimatel for their collaboration for joint project efforts and data gathering of this study. We are also grateful to Tekes, The Marcus Wallenberg Foundation, the Foundation for Economic Education, Yksityisyrittäjäin Säätiö and Metsämiesten Säätiö Foundation for their support.

## References

1. North, D, Smallbone, D. and Vickers, I. (2001). Public sector support for innovating SMEs. *Small Business Economics,* 16(2): 303–317.
2. Teece, D. J. (2007). Explicating dynamic capabilities: The nature and microfoundations of (sustainable) enterprise performance. *Strategic Management Journal*, 28(13): 1319–1350.
3. Schilling, M. A. (2008). *Strategic Management of Technological Innovation* (Boston, MA: McGraw-Hill Irwin).
4. Chiaroni, D., Chiesa, V. and Frattini, F. (2008). Patterns of collaboration along the bio-pharmaceutical innovation process. *Journal of Business Chemistry*, 5(1).
5. Branzei, O. and Vertinsky, I. (2006). Strategic pathways to product innovation capabilities in SMEs. *Journal of Business Venturing*, 21(1): 75–105.
6. Audretsch, D. B. (2004). Sustaining innovation and growth: Public policy support for entrepreneurship. *Industry and Innovation*, 11(3): 167–191.
7. Chesbrough, H. (2012). Open innovation: Where we've been and where we're going. *Research–Technology Management*, 55(4): 20–27.
8. Lee, S., Park, G., Yoon, B. and Park, J. (2010). Open innovation in SMEs — An intermediated network model. *Research Policy*, 39(2): 290–300.
9. Dijksterhuis, M. S., Van Den Bosch, F. A. J. and Volberda, H. W. (1999). Where do new organizational forms come from? Management logics as a source of coevolution. *Organization Science*, 10: 569–582.
10. Bergek, A. and Norrman, C. (2008). Incubator best practice: A framework. *Technovation*, 28(1): 20–28.
11. Chiaroni, D., Chiesa, V. and Frattini, F. (2011). The open innovation journey: How firms dynamically implement the emerging innovation management paradigm. *Technovation*, 31(1): 34–43.
12. Christensen, Jens Frøslev, Michael Holm Olsen and Jonas Sorth Kjær (2005). The industrial dynamics of open innovation — Evidence from the transformation of consumer electronics. *Research Policy*, 34: 10.
13. Von Hippel, E. (2005). *Democratizing Innovation* (MIT Press, Cambridge, MA).
14. Huizingh, E. K. (2011). Open innovation: State of the art and future perspectives. *Technovation*, 31(1): 2–9.
15. Nonaka, I. and Takeuchi, H. (1997). *The Knowledge-Creating Company*, 1995.
16. Hunt, S. D. and Morgan, R. M. (1995). The comparative advantage theory of competition. *Journal of Marketing*, 59(2).
17. Gassmann, O. (2006). Opening up the innovation process: Towards an agenda. R&D *Management*, 36(3): 223–228.
18. Crossan, M. M. and Apaydin, M. (2010). A multi dimensional framework of organizational innovation: A systematic review of the literature. *Journal of Management Studies*, 47(6): 1154–1191.

19. Teece, D. J. (1998). Capturing value from knowledge assets: The new economy, markets for know-how, and intagible assets. *California Management Review*, 40(3).
20. Ahlstrom, D. (2010). Innovation and growth: How business contributes to society. *The Academy of Management Perspectives*, 24(3): 11–24.
21. Kaplan, R. S. and Norton, D. P. (1996). *The Balanced Scorecard: Translating Strategy into Action*. Harvard Business Press.
22. Claycomb, C., Dröge, C. and Germain, R. (2001). Applied process knowledge and market performance: The moderating effect of environmental uncertainty. *Journal of Knowledge Management*, 5(3): 264–278.
23. Ford, D. and Saren, M. (2001). *Managing and Marketing Technology* (London: Cengage Learning Business Press).
24. Dodgson, M., Gann, D. and Salter, A. (2006). The role of technology in the shift towards open innovation: The case of Procter & Gamble. *R&D Management*, 36(3): 333–346.
25. Kirschbaum, R. (2005). Open innovation in practice. *Research–Technology Management*, 48(4).
26. Chesbrough, H. and Crowther, A. K. (2006). Beyond high tech: Early adopters of open innovation in other industries. *R&D Management*, 36(3): 2006.
27. Ranganathan, C. and Dhaliwal, J. S. (2001). A survey of business process reengineering practices in Singapore. *Information & Management*, 39(2): 125–134.
28. Trkman, P. (2010). The critical success factors of business process management. *International Journal of Information Management*, 30: 125–134.
29. Van de Ven, A. H. and Poole, M. S. (1995). Explaining development and change in organizations. *Academy of Management Review*, 20: 510–540.
30. Davenport, T. H. and Short, J. E. (1990). The new industrial engineering: information technology and business process redesign. *Sloan Management Review*, 31(4).
31. Scheer, A. W. and Nüttgens, M. (2000). *ARIS Architecture and Reference Models for Business Process Management* (pp. 376–389). Springer Berlin Heidelberg.
32. Aguilar-Saven, R. S. (2004). Business process modelling: Review and framework. *International Journal of Production Economics*, 90(2): 129–149.
33. Strnadl, C. F. (2006). Aligning business and it: The process-driven architecture model. *Information Systems Management*, 23(4): 67–77.
34. Costin, H. I. (1998). *Strategies for Quality Improvement* (2nd Ed.). The Dryden Press, Orlando.
35. Keller, S. and Price, C. (2011). Organizational health: The ultimate competitive advantage. *McKinsey Quarterly*, 2: 94–107.
36. Schroll, A. and Mild, A. (2011). Open innovation modes and the role of internal R&D: An empirical study on open innovation adoption in Europe. *European Journal of Innovation Management*, 14(4): 475–495.

37. Kajanus, M., Leskinen, P., Kurttila, M. and Kangas, J. (2012). Making use of MCDA methods in SWOT analysis — Lessons learnt in strategic natural resources management. *Forest Policy and Economics*, 20: 1–9.
38. Spence M. (2003). International strategy formation in small Canadian high-technology companies — A case study approach. *Journal of International Entrepreneurship*, 1(3): 277–296.
39. Eisenhardt, K. M. and Graebner, M. E. (2007). Theory building from cases: Opportunities and challenges. *Academy of Management Journal*, 50(1): 25–32.
40. Dodgson, M. (1993). Organizational learning: A review of some literatures. *Organization Studies*, 14(3): 375–394.
41. Antony, J. and Banuelas, R. (2001). A strategy for survival. *Manufacturing Engineer*, 80(3): 119–121.
42. Stephens, J. S. and McDonald Jr., C. L. (2010). Lean Six Sigma. The *Journal of Organizational Leadership and Business*. Retrieved November 24 (2007): 2010.
43. Evans, J. R. and Lindsay, W. M. (2005). *The Management and Control of Quality*, 6th Ed. (South-Western, Mason, OH).
44. Kwak, Y. H. and Anbari, F. T. (2006). Benefits, obstacles, and future of six sigma approach. *Technovation*, 26(5): 708–715.
45. Johnson, A. and Swisher, B. (2003). How six sigma improves R&D. *Research Technology Management*, 46(2): 12–15.
46. Archer, N. and Ghasemzadeh, F. (1999). An integrated framework for project portfolio selection. *International Journal of Project Management*, 17(4): 207–216.
47. Thore, S. A. (ed.) (2002). *Technology Commercialization: DEA and Related Analytical Methods for Evaluating the Use and Implementation of Technical Innovation* (Kluwer Academic Publishers, Boston).
48. Stummer, C. and Heidenberger, K. (2003). Interactive R&D portfolio analysis with project interdependencies and time profiles of multiple objectives. *IEEE Transactions on Engineering Management*, 50(2): 175–183.
49. Gustafsson, J. and Salo, A. (2005). Contingent Portfolio Programming for the Management of Risky Projects. *Operations Research*, November/December 2005, 53(6): 946–956.
50. Martino, J. P. (1995). *Research and Development Project Selection*. John Wiley and Sons, New York.
51. Liesiö, J., Mild, P. and Salo, A. (2007). Preference programming for robust portfolio modeling and project selection. *European Journal of Operational Research*, 181(3): 1488–1505.
52. Salo, A., Mild, P. and Pentikäinen, T. (2006). Exploring causal relationshis in an innovation program with robust portfolio modeling. *Technological Forecasting & Social Change*, 73: 1028–1044.

53. Belton, V. and Stewart, T. J. (2001). *Multiple Criteria Decision Analysis: An Integrated Approach*. Kluwer Academic Publishers, Dordrecht.
54. Ewing Jr., P. L., Tarntino, W. and Pamell, G. S. (2006). Use of decisioni analysis in the army base realignment and closure (BRAC) 2005 military value analysis. *Decision Analysis*, 3: 33–49.
55. Mumford, M. D. and Licuanan, B. (2004). Leading for innovation: Conclusions, issues, and directions. *Leadership Quarterly*, 15: 163–171.
56. Von Hippel, E. and von Krogh, G. (2006). Free revealing and the private-collective model for innovation incentives. *R&D Management*, 36(3): 295–306.
57. Seppänen, M., Pellikka, J. and Littunen, H. (2014). *Inter-Organizational Collaboration, Sources of Innovation, and a Firm's Performance*. This paper was presented at The XXV ISPIM Conference — Innovation for Sustainable Economy & Society, Dublin, Ireland on 8–11 June 2014.
58. Birkinshaw, J., Hamel, G. and Mol, M. J. (2008). Management innovation. *Academy of Management Review*, 33: 825–845.
59. Pellikka, J. (2014). "The Commercialization Process of Innovation in Small High-Technology Firms — Theoretical Review." *Handbook of Research on Techno-Entrepreneurship: How Technology and Entrepreneurship are Shaping the Development of Industries and Companies*, pp. 91–108.
60. Vaccaro, I. G., Jansen, J. J., Van Den Bosch, F. A. and Volberda, H. W. (2012). Management innovation and leadership: The moderating role of organizational size. *Journal of Management Studies*, 49(1): 28–51.
61. Chesbrough, H. and Schwartz, K. (2007). Innovating business models with co-development partnership. *Research–Technology Management*, 50(1): 55–59.

# Chapter 9

# OPEN INNOVATION NETWORKS: EXPLORING ACTOR ROLES AND NETWORK ORCHESTRATION IN LIVING LAB

Dimitri Schuurman,* Bram Lievens,† Carina Veeckman,† Lieven De Marez* and Pieter Ballon†

**iMinds — MICT — Ghent University*
*Korte Meer 7-9-11, 9000 Ghent, Belgium*
*†iMinds — SMIT — VUB, Pleinlaan 9*
*1050 Brussels, Belgium*

In our contemporary society, innovation is more and more considered as a collaborative effort between different actors who tap into distributed sources of knowledge. This has fostered research into open innovation, innovation networks and open innovation processes. However, research on how these innovation processes are coordinated within these networks is largely lacking. Therefore, within this chapter we explore a specific approach that tries to facilitate and govern distributed innovation processes through a Public–Private–People partnership with a central role for the end-user: Living Labs. As we consider Living Labs as clear examples of distributed innovation through a collaborative effort of different actors, we use concepts from open innovation and from innovation networks literature to analyse and make sense of innovation processes taking place in Living Labs. By means of an in-depth case study analysis, we look at the knowledge transfers, constellation and orchestration taking place in FLELLAP, a Flemish ICT Living Lab that generated some successful projects, but also failed on some aspects. This allows to abstract lessons and propose avenues for future research.

*Keywords*: Open Innovation; Innovation Networks; Living Labs; Systemic Instruments; Distributed Innovation; Co-Creation.

## 1. Introduction

Since the industrial revolution started near the end of the 18th century, and especially with the advent of the so-called Fordism at the beginning of the 20th century, the dominant production and innovation logic aimed at vertical integration within the boundaries of a firm or company.[1] Only near the end of the 20th century, in terms of innovation, this dominant view was challenged in favour of a more distributed view. This shift in the dominant mode of innovation, from vertically integrated innovation towards a more distributed mode of innovation, has forced companies to alter both their research and development processes and their approach to innovation management. Instead of focusing on hiring people with all relevant skills and knowledge, and investing heavily in internal research and development capacities, companies had to actively look outside for knowledge and technology to complement internal assets. This shift in the dominant mode of innovation not only required companies to adapt by developing or acquiring different skills and abilities, it also encouraged a growing body of research into the nature and occurrence of distributed innovation processes.[2] In academic theory, this shift has already taken place, but in practice, a lot of companies and innovation practitioners are still struggling with the concrete implementation of strategies to cope with these distributed innovation processes.[3] Within this chapter, we explore a specific approach that tries to facilitate and govern distributed innovation processes through a Public–Private–People partnership with a central role for the end-user: Living Labs. We consider Living Labs as clear examples of distributed innovation through a collaborative effort of different actors, therefore we study Living Labs through the lens of Open Innovation and innovation networks in order to make sense of innovation processes taking place inside Living Labs. By means of an in-depth case study analysis, we will look at the knowledge transfers, constellation and orchestration taking place from a network perspective.

## 2. A Process Perspective on Open Innovation

Already one year after Chesbrough's[4] initial book on Open Innovation, in which the main message was that opening the internal innovation process of a firm yields extra value, Gassmann and Enkel[5] further explored this rapidly emerging concept and took a process perspective on Open Innovation. Based on a case study of the IBM Industry Solution Lab in Zürich, they demonstrated that the locus of the various innovation activities is decoupled into three parts. First, there is the locus of knowledge creation, the locus of innovation (indicating applying the idea/knowledge/technology and transforming it into an innovation), and the locus of commercialisation (product development or exploitation of the innovation). The fact that the locus of innovation shifts during the innovation development process implies the existence of *knowledge transfers*. Moreover, by distinguishing a locus of knowledge creation, a knowledge of innovation and a locus of commercialisation, Open Innovation also acknowledges the fact that innovation is a process.[6]

Research into these transfers lead to the identification of three core Open Innovation processes: **outside-in** (enriching the company's knowledge base through integrating customers and suppliers and through external knowledge sourcing), **inside-out** (getting pecuniary returns for transferring ideas/knowledge/technologies to the outside environment), and the **coupled process**, which combines both inside-out and outside-in processes by working in alliances with complementary partners.[5] West and Bogers[7] argue that the outside-in or inbound process has received most research attention and ascribe this to the fact that this process builds further on a large body of prior research. They see three different research angles on this inbound process: How firms obtain external innovation, the role of innovation created outside the firm by individuals, and research on open source software (including open source communities).

Instead of outside-in and inside-out, **exploration** and **exploitation** are also sometimes used.[6] These are concepts introduced in the context of organisational learning and refer to the relation between the exploration of new possibilities and the exploitation of old certainties.[8] Purposive outflows of knowledge or knowledge exploitation, implies innovation activities to leverage existing technological capabilities outside the boundaries of the organisation, whereas purposive inflows, or knowledge exploration, relates to innovation activities to capture and benefit from external sources of knowledge to enhance current technological developments.[9] The coupled process can then be seen as a form of simultaneous exploration and exploitation between two companies or organisations. Interestingly, Enkel *et al.*[10] equal the coupled process to **co-creation with complementary partners** that engage in simultaneous outside-in and inside-out processes through alliances, cooperation and/or joint ventures.

## 3. Open Innovation Systems, Networks and Systemic Instruments

Most of the Open Innovation concepts and literature takes a single firm as the unit of analysis or the exchange process of knowledge between firms. However, these exchanges do not necessarily occur between two actors, but can also be a part of a larger network or constellation of actors engaging in Open Innovation. As Living Labs consist of multiple actors engaging in innovation projects that deal with knowledge transfers and collaborative activities (such as co-reation), we need to take a more encompassing perspective that is able to analyse and describe these networked structures and activities. Therefore, we turn to the literature on systemic innovation, (open) innovation networks and systemic instruments for concepts and frameworks to take into account this network view on Living Labs.

The first concept, we introduce has received the least attention, especially in terms of empirical research, and operates on a more 'macro' level compared to innovation networks: **innovation systems**. Although there is no consistent definition available (yet), the concept of the innovation system sees technology and information transfers between people, companies, organisations and institutions as essential to the innovation process. Wieczorek and Hekkert[11] state that, in its broadest definition, an innovation system entails all parts and aspects of an economic structure, together with the institutional set-up affecting learning, searching and exploring, which includes the production, marketing and finance system. Actors within innovation systems include users, producers, intermediary organisations and supportive organisations.[12] Wieczorek and Hekkert[11] discern between different types of innovation systems: national or regional innovation systems (when a geospace is a unit of analysis), sectorial innovation systems (dealing with a whole sector of economic activity, often going beyond national borders), and technological innovation systems (evolving around a specific technology). This means that innovation systems consist of multiple levels that interact with each other and that can be subjected to analysis. Fichter[13] discerns the company level (the most used level of analysis in Open Innovation research), the value chain level of innovating actors, and the level of framing and interlinking organisations (a superstructure level that hosts organisations that enable other actors to innovate).

Following the broad definition of innovation systems, we can see **innovation networks** as part of these systems, where innovation networks can be defined as purposefully established connections between the demand-side (intermediate and end-users of innovation) and the supply-side (producers of knowledge and technologies) of the knowledge infrastructure, as well as with other relevant actors from within the innovation system.[14] Within this system and network perspective co-operation between several different types of actors is seen as a key to successful innovation. Fischer[15] relates this to the belief that innovation networks offer time advantages over internal development in realising innovations in a shorter time interval. Rese and Baier[16] discovered the following four network-related success factors that enhance the chances of successful innovation in innovation networks: *Trust*, *commitment*, *dependency*, and *compatibility* of the network actors. Referring back to the Open Innovation processes of exploration and exploitation, Dittrich and Duysters[17] hypothesise that actors pursuing an exploration strategy will look for partners with distinctly different capabilities, resulting in an innovation network consisting of partners in new technological areas, whereas companies following an exploitation strategy will search for companies with similar technological capabilities, resulting in an innovation network of partners in similar technological areas.

Rese and Baier[16] state that innovation networks appeal in particular to SMEs as they are able to reduce existing barriers for innovation by complementing the resources they lack. This resonates with the hypothesis put forward by Westerlund and Leminen[18] that small companies find it easier to switch to an Open Innovation strategy (cf. *supra*).

However, research into innovation systems and networks also revealed that several gaps might hinder effective co-operation. Based on an in-depth case study of an innovation network in the Dutch agriculture industry, Klerkx and Aarts[19] provide an overview of the three main challenges and paradoxes that occur when dealing with open innovation networks. First, they need to find a balance between new relationships and existing relationships, i.e., balancing openness (exploiting weak ties) and closure of the network (fostering strong ties). This is also referred to as the need for 'dynamic stability'.[20] Second, the ways of interaction between the actors in the network need to take into account the different perspectives of the actors on the nature and form of the cooperation, as (too) diverging perspectives may lead to conflicts or even network failure. Third, the formal and informal relationships between the actors in the network need to be balanced. Actors in the network need to manage the paradox that they have to develop their position in the network to reap the benefits, but that total control of the network by one firm may be counterproductive as it undermines the informal basis of network cooperation.

These challenges can be overcome through the process of network orchestration. The literature regarding this orchestration has identified three basic elements[19,21,22]: **Demand articulation**, which refers to the continuous vision development and articulation of related technology, knowledge, and other resource needs; **innovation network composition**, which entails scanning, filtering, and matchmaking of new network partners for accessing certain resources; and **innovation process management**, or coordinating the network to foster trust, transparency, and reciprocity.

This led to the creation of **systemic instruments** that aim to address problems, referred to as systemic weaknesses or systemic failures, that arise at the innovation system level and which negatively influence the speed and direction of innovation processes that run in the innovation networks within the innovation system.[12] One of these 'instruments', advocated by scientific and policy literature, are so-called intermediary organisations to fulfill bridging and brokerage roles within innovation systems to overcome various gaps among innovation system stakeholders that can lead to innovation system failures and reduced performance.[14] Katzy *et al.*[23] also state that there is broad agreement in literature that innovation processes in open networks are coordinated through a visible hand, often referred to as **innovation intermediary**, and propose the Living Lab as a process coordinating innovation intermediary for '(1) closing the pre-commercial gap by manifesting initial demand for products and services, as well as (2) orchestrating the actions of disparate actors in order to gain critical mass for the creation of a product or service'.[24] These innovation intermediaries are described to provide a set of operative activities that link them to the network innovation processes, but literature provides only fragmented insight about the intermediary–process relationship. Sieg *et al.*[25] also note that there is a relatively limited understanding of the implementation of open innovation through innovation intermediaries, with the notable exceptions of knowledge brokers (recombination of existing solutions)[26,27], and virtual knowledge brokers.[28] Moreover, Smits and Kuhlmann[12] note that the most systemic instruments, such as brokers and intermediaries, take an individual organisation or a bilateral relation as unit of analysis, focusing largely on the private sector and far less on the public sector and public–private alliances, with only very less attention for learning processes, platforms for experimentation or tailor-made strategic intelligence.

This apparent gap in the literature is somewhat filled by the research stream focusing on Triple and Quadruple Helix models. We already introduced the Triple Helix model when discussing the System Models of Innovation Management (cf. *supra*). The Triple Helix thesis states that the university can play an enhanced role in innovation in increasingly knowledge-based societies[29], and represents an evolutionary model of innovation that looks at the interactions between three distinct actors ('helices'): university, industry and government.[30] By acknowledging the interactions between and influence of these actors on one another, the Triple Helix as an analytical model looks to describe institutional arrangements and policy models, conceptualising innovation as inherently dynamic and driven by various forces interacting with each other[29], which resonates with the evolutionary models of innovation. A connection between these systemic and evolutionary models on innovation and Living Labs is made by introducing Quadruple Helix models that add the user as a fourth helix. Arnkil *et al.*[31] see the evolution towards Quadruple Helix models as an acknowledgement of broad cooperation in innovation, and a shift towards systemic, open and user-centric innovation policy, as opposed to an era of linear, top–down, expert driven development, production and services.

Katzy *et al.*[23] consider Living Labs as open innovation networks that function as systemic instruments or innovation intermediaries that try to overcome innovation barriers for the involved actors in the Living Lab network. However, research on how these innovation processes are coordinated is largely lacking, especially literature linking a more systemic view with the open innovation processes.

## *Living Labs: From user research towards open innovation networks*

If we consider the European innovation system, there is an apparent strength in generating knowledge, while the translation into actual successful innovations is lagging behind. This is referred to as the 'European Paradox' or the gap between research leadership and the commercial success of innovation.[32,33] Almirall and Wareham[24] rephrased this 'European Paradox' in terms of open innovation concepts and stated that Europe scores high in terms of research (= exploration), but underperforms in terms of market success (= exploitation). In order to overcome this paradox, several initiatives were started at the European policy level, such as the promotion and support of industry–university links and relationships.[34] A specific case of industry–university relationships received considerable support from Europe: Living Labs.[35] In the 1990's, the concept of Living Labs already appeared in academic discussions, but this policy support by the European Commission in 2006, stimulating projects to advance, coordinate and promote a common European innovation system based on Living Labs, provided a boost to Living Lab practice.[36] The most noteworthy policy support came with the establishment of the European Network of Living Labs (ENoLL), an organisation aimed at connecting Living Labs for knowledge exchange, networking purposes and the development of a shared innovation concept with to date, according to their website, more than 300 Living Labs being linked to the ENoLL, mainly in Europe but also in the rest of the world.[37]

One of the first appearance of Living Labs in the European academic literature pops up in the discussions regarding the impact of technology in general, and ICT in particular, on society. Frissen and van Lieshout[38] define Living Labs as consciously constructed social environments in which the uncontrollable dynamics of everyday life are accepted as part of the innovation environment which enables designers and users to co-produce new products and services. This first and oldest definition from our review focuses on user involvement and on the everyday context as an important divergence from more traditional views on innovation. Living Labs are seen as research approach to study the impact of technology on society, but also the impact of society on technology, which links this to the mutual shaping-perspective from the social sciences. Ballon *et al.*[39] define Living Labs as an experimentation environment in which technology is given shape in real life contexts and in which (end-)users are considered 'co-producers', and situate them amongst other Test and Experimentation Platforms with the maturity of the innovation perceived as medium and with a focus between designing and testing. Eriksson *et al.*[40] and Niitamo *et al.*[41] see Living Labs as ecosystems fostering user-centered innovation through experimental platforms where the users are studied in their everyday habitat, and this by means of quantitative as well as qualitative research methods with the focus on accessing the ideas and knowledge of these users. The user-centric aspect is once again very dominant, as well as the everyday habitat, but the co-creative aspect is not explicitly mentioned here. A multi-methodical research approach is also put forward as inherent in Living Lab-practice. According to Kusiak[42], the Living Lab concept calls for all stakeholders of a product or a service to participate in the development process. It supports innovation of products and services that are validated in collaborative, multi-contextual, empirical real-world environments. The main difference between the traditional consumer research programs and the Living Lab approach is in the multi-role and multi-faceted involvement of the customer. Living Labs as some kind of aggregator of various external inputs, translating them into requirements, the Living Innovation Laboratory. The Living Lab is considered as some kind of innovation intermediary.

Schaffers *et al.*[43] define Living Labs as user-centric environments for open innovation characterised by early and continuous involvement of users and by user-driven rapid prototyping cycles. Here, the focus shifts towards the front-end of innovation (the early involvement) and iteration of the innovation in development. Ståhlbröst and Bergvall-Kåreborn[44] give a quite similar definition, but again mention co-creation explicitly as they state that Living Labs are a means to gain access to the ideas, experiences, and knowledge that users possess, built upon co-operation with users to support creativity, so an efficient interaction with a larger population of people should be facilitated. They place Living Labs within a strong user-centric approach but do not stress the everyday habitat. Levén and Holmström[45] also put forward co-creation as the central process permeating all Living Lab operations, with a focus on the consumer, but also in the context of a public–private partnership. Feurstein *et al.*[46] see Living Labs as a systemic innovation approach in which all stakeholders in a product, service or application participate directly in the development. Living Labs are thus seen as collaborations of public–private–civic partnerships in which stakeholders co-create new products, services, businesses and technologies in real life environments and virtual networks in multi-contextual spheres. This definition stresses the multi-stakeholder aspect of Living Labs with less exclusive attention for the end-user, who is only one of the stakeholders involved. This also relates to the Living Labs aimed at collaboration and knowledge exchange.[47] Turkama[47] distinguishes the underlying principles of Living Labs as opposed to those from the closed innovation model and user driven Open Innovation. Here, Living Labs are seen as an extension of the principles of Open Innovation. According to Turkama[47], the main principles of Living Labs are:

— Crowdsourcing the people's wisdom leads to smarter products/services.
— R&D and innovation must be done with the users in open, real-life testing environments.
— Time-to-market & market potential are enhanced by working with users.

— Social and institutional factors are as important as technology and economics in driving innovation.
— New and better ideas come from users' integration since the very early stages of product/service development.
— A transparent public–private partnership is the way to deploy innovation and to achieve more sustainable results in the medium-to-long run.

Besides the principles we also abstracted from most of the previous definitions (a public–private partnership, real-life contexts, early-stage user involvement), there are also some links to the methodological level, as 'crowdsourcing' is explicitly mentioned, and to the eventual outcomes of Living Labs, a shorter time-to-market and a higher market potential. Summarising, these principles are rather diffuse and deal with methodological aspect, organisational aspects and outcomes of Living Labs without making a clear distinction between them.

The final definition in this time-based overview comes from Almirall and Wareham[24] who state that "Living Labs are semi-partitioned spaces in the form of innovation arenas integrated in real-life environments but separated by means of an innovation project structure that cultivate user-led insights" and "Living Labs are fundamentally infrastructures that surface tacit, experiential and domain-based knowledge such that it can be further codified and communicated". Bearing in mind the 'European Paradox', it is surprising that 'exploitation' is not mentioned explicitly in the definitions, whereas exploration is. We also wish to stress the 'innovation project structure' mentioned by Almirall and Wareham[24], as we see this as an important distinguishing element in Living Labs. A Living Lab innovation project is carried out within a given Living Lab infrastructure, but in most literature this distinction is not explicitly made, as both are mostly referred to as 'Living Labs'. Based on the findings from Schuurman[48] and from the evolution of Living Lab definitions, we propose the following sublevels to be distinguished in Living Labs: A macro level, consisting of the Living Lab actors that compose the innovation network, a meso level which consists of the

Living Lab projects carried out within the constellation, and the micro level which consists of the methods that are being used in the Living Lab projects. Within this chapter, our analysis is focussed on the macro level.

Within these overview of definitions and conceptualisations we can see an evolution from Living Labs as user centered and user driven approaches towards a more fully eco-system driven approach, with attention for all (possible) stakeholders in the innovation process. Whereas in the earlier definitions the 'innovation arena' is situated in society as a whole, we see more realism in later definitions, as is the case with Almirall who distinguishes the Living Lab by a project-based space. Later definitions also acknowledge the diversity of approaches and evolve into less 'strict' definitions and more in terms of elements, 'philosophies' or principles that can or should be present in Living Labs. The natural context, albeit artificial or real-world, is also a constant in these definitions, as well as the conceptualisation of innovation as a process consisting of various steps.

Leminen *et al.*[49] base most of their published research on this matter on an analysis of 26 Living Labs from Finland, South Africa, Spain and Sweden. They obtained data form these Living Labs by conducting 103 semi-structured interviews with informants from 39 different organisations participating in these Living Labs. The interviewees included senior managers, project managers, researchers, project coordinators and users. Within the various publications that are based on these data, they take different perspectives which enable to draw a clearer picture of the complexity of Living Labs, their actors and their roles.

Based on an empirical investigation of multiple Living Labs, Leminen *et al.*[49] propose four different Living Lab actors based on their role: utilisers, enablers, providers and users.

- ***Utilisers*** aim to develop their businesses within the Living lab ecosystem, mostly through short-term Living Lab cases. Their focus is on developing and testing their new products and services. These utilisers use Living Labs as a strategic tool to collect data on test-users of their products or services and collaborate with all stakeholders in the Living Lab ecosystem, including the end-users. These actors drive short-term Living Lab projects and can be regarded as short-term, *ad hoc* 'consumers or partners of the Living Lab'.

- ***Enablers*** can be various public sector actors, non-governmental organisations, or financiers, such as towns, municipalities, or development organisations. This actor provides (financial) resources or policy support in order to start-up and maintain the Living Lab operations.
- ***Providers*** provide other actors in the Living Lab with their product or service portfolio. They take care of the (material) infrastructure used for the Living Lab operations. Providers are mainly private companies that enter into Living Labs to co-develop new products, services, and solutions to their own business or industry needs, and focus more on long-term results. They attain these goals through their involvement in general Living Lab operations and (possibly) in the Living Lab cases, driven by utilisers.
- ***Users*** are the 'end-users' that are being involved in the Living Lab-operations and in the (short-term) Living Lab cases. In some Living Labs, existing user groups or user communities are involved, while in others the Living Lab operations themselves facilitate the formation of a Living Lab user community.

In the typology of Leminen *et al.*[49] academic researchers are considered providers because they provide the necessary expertise on user research. Other research such as the triple and quadruple helix concepts, however, stresses the importance of universities as a distinct actor in the innovation ecosystem.[31,34,50,51] Moreover, the contribution of academia is not limited to user research, as it can also include research on technical topics related to the focus of the Living Lab or policy and business researchers. Therefore, we distinguish ***researchers*** as a separate type of actor within the Living Lab anatomy.

Based on the various roles of the Living Lab actors and the central role of the infrastructure, we propose the following theoretical model of a Living Lab.

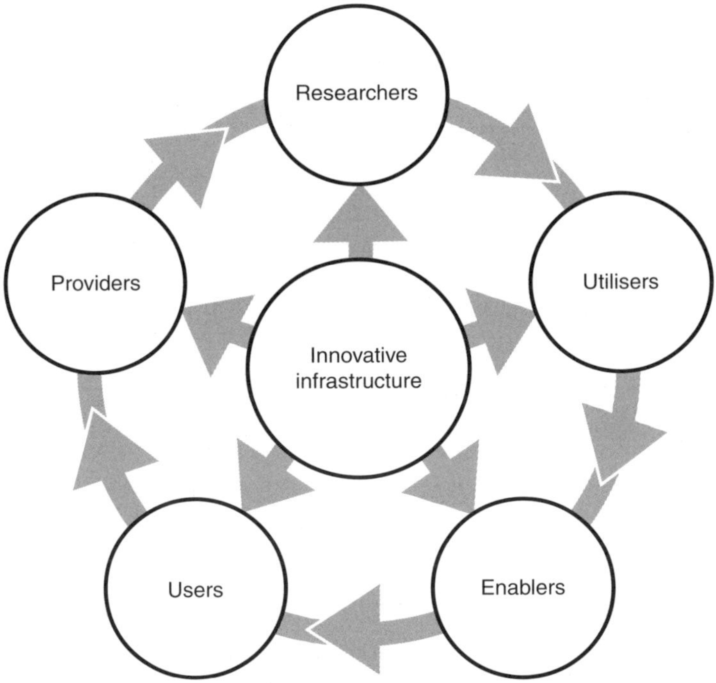


## 4. Methodology & Hypotheses

As an empirical data gathering and analysing technique, we used the case study technique, a common method in social sciences to describe and explore poorly understood processes and events. Case studies are especially suited because of their emphasis on detailed contextual analysis of a limited number of events or conditions and their relationships.[52] Yin[53] defines the case study research method as an empirical inquiry that investigates a contemporary phenomenon within its real-life context; when the boundaries between phenomenon and context are not clearly evident, and in which multiple sources of evidence are used. Given the complexity of the studied phenomenon, the multiple levels of analysis (innovation network, Living Lab project, knowledge flows, etc.), this research design seems most appropriate. Finally, our study also shows some elements of an action research design[54] as members of the author team participated themselves in the cases studied by researchers, embedded in real projects and interacting in real-life settings with the different actors participating in the Living Labs and in the Living Lab projects in order to help solving problems and learn from this experience.[55] This position provided the authors with in-depth, rich insights and access to all sorts of data sources. This unique position allowed to constantly shift between reflection and theorising on the one hand, and actively doing and putting into practice on the other hand.

For our analysis, we were able to use the following data sources as first-hand involved actor in the Living Lab operations and Living Lab cases:

— Official meeting minutes of all steering committees and of all official work package meetings.
— The initial project proposals and all project reports.
— All deliverables from the Living Lab operations and of the Living Lab cases.
— All data from user research regarding Living Lab operations (intake surveys, domestication interviews, ...) and regarding the Living Lab cases.
— Field notes of all Living Lab cases meetings.
— Data from a short survey that was held amongst all consortium partners from LeYLab at the end of the Living Lab, which took the form of a SWOT-exercise.
— In depth interviews with the three Living Lab managers from FLELLAP, LeYLab and Mediatuin.
— Semi-structured interviews with the instigators from all the Living Lab projects from the analysis.

The interview guides were pretested with colleague Living Lab researchers from iLab.o. The results were also discussed with them. For the analysis of all interviews, we used the technique of affinity diagramming, which originates from the User Centered Design tradition[56] and allows us to discover emerging themes and topics in the research data. We did this by extracting quotes from the transcribed interview and grouping quotes that dealt with similar topics.

From the literature on open innovation and from our Living Lab actor model, we abstract the following hypotheses:

For **utilisers,** we expect exploration as main motive in order to stimulate their innovation processes. The Living Lab provides the opportunity to get need information and solution information from the users involved in the Living Lab. For them, it is an ecosystem in which they can develop, test and learn.

The **researchers** are expected to function as intermediaries between utilisers and users, as through their research they are able to abstract the need and/or solution information from the users, which the utilisers are looking to explore. However, the Living Lab operations and activities also allow researchers to explore their own knowledge base (testing hypotheses, generating new theories/methodologies, etc.). They expect to generate research data that can be academically valorised. By doing so, researchers contribute to the knowledge retention of the Living Lab.

For **providers,** we expect exploitation of the technology and/or knowledge they bring into the Living Lab network as main motive. They expect the Living Lab operations to provide them with input for their market strategy and roadmap. On top of that, the Living Lab enables them to showcase their innovative infrastructure.

For **users**, we expect intrinsic motivations (such as task enjoyment and curiosity) to participate in the Living Lab to be dominant, as knowledge transfers between users and producers have found to be non-pecuniary. However, extrinsic motivations might also play a role (incentives and social value).

As **enablers** contribute to the Living Lab with money or others assets that enable the Living Lab operations, these public organisations expect the Living Lab to fulfill some predefined policy goals. Mostly, this concerns the generation of social and/or economic value, such as increased neighbourhood cohesion or stimulating innovation and entrepreneurship.

| | Motivations to participate in the Living Lab | Value created by the Living Lab |
|---|---|---|
| Utilisers | Develop, test and learn<br>Exploration for innovation | Need information<br>Evaluative information<br>Solution information |
| Enablers | Meet policy goals | Economic value creation<br>Social value creation |
| Providers | Exploitation of provided infrastructure<br>Exploration of market introduction | Pecuniary returns<br>Showcase & test infrastructure |
| Users | Intrinsic motivations over extrinsic motivations | Fun<br>Knowledge<br>Empowerment |
| Researchers | Exploiting implementable knowledge<br>Exploring new knowledge | Pecuniary returns<br>Research data for valorisation |

In the proposed conceptual model of a Living Lab, the **infrastructure** has a central role since it facilitates collaboration among all actors and enables knowledge and technology spill-overs within the innovation ecosystem.

## 5. Results

### *Flemish Living Lab Platform — Vlaams Proeftuin Platform (FLELLAP)*

FLELLAP officially started in October 2010, to support the development of innovative information, communication and entertainment (ICE) products and services.[a] Its mission was to boost the valorisation of ICE research and development in Flanders and to support joint value creation for all involved stakeholders. FLELLAP is a consortium of several industrial partners (Telenet, FifthPlay, Androme & Alcatel-Lucent) and the **iMinds** department **iLab.o**.

[a] http://vlaamsproeftuinplatform.be/en/about-us/.

**Telenet** (www.telenet.be), one of the major telecom operators of Flanders, was coordinating the initiative. It is a traditional telecom operator who delivers cable television, high-speed Internet access and fixed and mobile telephony services, mainly via their cable network. **Androme** (www.androme.be/) and **FifthPlay** (http://www.fifthplay.com/) are two innovative SMEs. The former offers high-quality software solutions to large multinationals and SMEs, the latter develops and produces innovative technologies for energy management, health monitoring and builds upon the integration of technologies in buildings and cities and is 100% owned by Niko[b], a hardware company. The last named industrial partner, Alcatel-Lucent (www.alcatel-lucent.com), is a worldwide solution and service provider, with the prominent research and development center Bell Labs.

Initially, as was stated in the project proposal, the Living Lab focused on three domains: Smart Cities (FifthPlay), Smart Grids (Alcatel-Lucent) and Smart Media (Telenet), with each partner dedicated to one field and Androme providing technical support where necessary. This way FLELLAP would consist of three Living Lab constellations with each its dedicated infrastructure and a separate panel of test users. However, due to multiple reasons, that will be dealt with later on, this ambition was never realised. No active Smart Grids or Smart Media panel was recruited, while the Smart Cities panel was much smaller than predetermined (75 households and 250 end-users). This was mainly because of the lack of clearly defined internal Living Lab projects and also because of shifting goals and priorities among the consortium partners.

For the Smart Media theme, Telenet had initially foreseen to roll out 100 3D TVs for user testing with content and various applications, but after FLELLAP had started, it was decided internally that 3D was no longer a priority for Telenet, something which was announced during the first steering committee meeting by a Telenet representative. Eventually, Telenet did an in-house user test with 3D TV, but the size of this project was much smaller than planned and this

[b] http://www.niko.eu/nlnl/niko/over-niko/niko-group/.

took place only during the later stages of FLELLAP. As became clear later on in the project (cf. *infra*), their focus had turned towards other television services, such as the development of a second screen streaming application, Yelo TV, together with Androme.[c] This project was carried out outside the scope of FLELLAP, but resulted in a small scale user test with FLELLAP panel members, but this also took place in 2013, near the end of the Living Lab.

Alcatel-Lucent intended to integrate its testing laboratory for energy monitoring applications, but around the time FLELLAP started, an internal company re-orientation caused less commitment towards exploiting this technical lab. As for the Smart City theme, FifthPlay took care of 50 tablets and gateways that were preconfigured to run the InCitys platform that was developed by FifthPlay, but in the meantime FifthPlay had also enabled a connection of the platform with a smart plug from its product portfolio which enabled to monitor the energy consumption. This way, FifthPlay also partly covered the Smart Grids topic, which potentially interfered with the leading role of Alcatel-Lucent within this thematic domain (although their focus was exclusively on B2B-applications). In terms of the infrastructure roll-out, only the FifthPlay project needed users, for which 250 people were recruited in the City of Sint-Niklaas, of which 50 were eventually equipped, as it was the intention to roll out extra services from (local) external utilisers on the platform. Telenet also acted as a provider by offering a specific modem (DOCSIS 3.0 modem) and free internet to the participants of the field trial. It is worth mentioning that the tablets were preconfigured to run the InCitys platform and could not be used for other applications, something which was not the case in LeYLab (cf. *infra*).

As in terms of panel recruitment and user research not much activity was planned, the researchers and the panel managers decided to build a larger panel that could be used when external utilisers would come to the Living Lab to initiate a Living Lab project. This was realised by conducting surveys during fixed time intervals that

[c] http://www.androme.com/index.php/news/35/70/Telenet-lanceert-digitale-tv-over-wifi.

covered different aspects for the three thematic domains. This way, one larger panel of 2.015 users was built up, allowing for data gathering and profiling of this user panel for external Living Lab projects. This way, FLELLAP repositioned itself towards a panel-based Living Lab in which the end-user became the major asset. Besides the smaller Telenet projects with 3D TV and YeloTV, this panel was used in three external cases in the Smart Media domain: Fietsnet, MuFoLive and OKEEZ.

The internal Telenet cases are not covered in our case analysis as the results and tests were for the most part an internal Telenet matter. The Okeez case is also not covered because we had no first-hand experience in this case and because it turned out to be not possible to schedule an interview because the company had moved to the United States. The other three cases, InCityS, Fietsnet and MuFoLive, will be covered in the project analysis.

FLELLAP applied for membership during the fifth wave and is still listed as one of the 21 effective members to this date, despite the fact that the Living Lab constellation ceased to exist in the course of 2013.

Analysis of actor roles

When analysing the Living Lab constellation according to the actor roles from Leminen and Westerlund[57], Telenet clearly is in the role of provider of the infrastructure, where the initial goal was to explore future applications and features for its cable network and look for additional services that could be offered to its customers, with a focus on Smart Media. Based on the project proposal, Alcatel-Lucent was also a provider within the Living Lab constellation, but due to a changed course from the management of the company, the Smart Grid laboratory for testing applications and services was reoriented towards B2B applications, so the laboratory infrastructure was never used for external utilisers. The goal of Alcatel-Lucent was initially to exploit the laboratory to utilisers and to explore the possibilities, but these goals were not realised. Androme can also be labelled as a provider as it offers its technical knowhow to the Living Lab constellation and was looking to exploit this knowledge in other projects. However, in practice, Androme only delivered services to Telenet, especially for the user interface of YeloTV, the second screen application that Telenet announced and launched during the running time of the Living Lab. These exchanges took place in a pecuniary modus, where Androme exploited its knowledge to Telenet. Androme also became a utiliser during the project as it did a user test with a user interface for video conferencing that they were developing at that time, which was not successful at all as they experienced lot of technical problems. The panel management from iLab.o is the fourth provider in the Living Lab constellation, providing the communication with and handling of the user panel and of the devices that were given to the users, and also providing the Living Lab constellation with the LLADA-tool (Living Lab Data Aggregator for storing all user information). Through the Living Lab operations, iLab.o's panel managers could further develop their skills and expertise, and additional users were recruited who increased the number of total available end-users for other projects.

FifthPlay is situated somewhere between the provider role and the utiliser role within the Living Lab constellation. FifthPlay wanted to explore the possibilities of its platform and used the Telenet infrastructure and the user panel recruited by the iMinds researchers and panel managers to roll-out their platform, but they also provided the gateway and the platform itself on dedicated tablets on which other services could also run. Attracting other service providers on the platform is a form of exploitation, although they were not very successful in this matter as local retailers and organisations were not very keen to get on the platform. After FLELLAP FifthPlay launched a commercial version of the platform on the market called Nuvonet[d], but according to FifthPlay themselves this has not been a huge success. One of the services that did run on the platform during the Living Lab, and which appeared to be one of the most valued applications by the test-users, was the energy monitoring application with smart electricity plugs from Niko. In the course of FLELLAP, FifthPlay was able to exploit this technology to Electrabel (www.electrabel.be) in the form of a joint 'smart boxes' offering that was launched as a commercial service to end-users.[e]

---

[d] http://www.nuvonet.be/en.

[e] http://www.fifthplay.com/en/news/press/smart-energy-box.

In the 3D TV and YeloTV projects, Telenet also acted as a utiliser within the Living Lab, but these cases did not seem to have had a high priority. For 3D TV this became apparent during the first steering committee and also due to the fact that the only 3D only channel, High 3D TV[f], was recently removed from the offering.[g] For YeloTV, most of the development was done outside of the Living Lab in alliance with Androme[h], and the user test only took place in the period between the sneak preview in October 2012[i] and the official launch in March 2013[j], which did not allow to include a lot of modifications. YeloTV is considered as a success by Telenet.[k]

It is striking that the consortium did not include any actor that could be labeled as purely a utiliser, which indicates that FLELLAP was conceived as a Living Lab that would be able to quickly attract external utilisers, but this turned out differently. This also forced the researchers in FLELLAP, who came from iMinds and were connected to iLab.o, to be creative and initiate research with the panel members in order to activate them. These users consisted of a separate 'Smart City'-panel of testers of InCityS, all living in the Flemish City Sint-Niklaas, and a user panel with 2.015 users that were recruited by iLab.o panel management and that were sent a survey once every 2 months regarding the three thematic domains. However, this also allowed to conduct some experiments regarding user motivations to participate and regarding panel retention (exploration of new knowledge), that resulted in research valorisation through some publications.[58–61]

As enabler of the Living Lab constellation, we can discern IWT (Institute for the Promotion of Innovation by Science and Technology in Flanders), a Flemish government agency, together with iMinds.

---

[f] http://www.hightv3d.com/.

[g] http://nl.wikipedia.org/wiki/Telenet_Digital_TV.

[h] http://www.uhasselt.be/UH/techtransfer/Voor-bedrijven/tevreden-klanten/tevreden-klanten-Androme-stond-mee-aan-de-wieg-van-YELO-TV.html.

[i] http://corporate.telenet.be/en/news-and-media/press-announcements/yelo-tv-brand-new-television-experience.

[j] http://corporate.telenet.be/en/news-and-media/press-announcements/telenet-launches-new-tv-experience-yelo-tv-5-march-2013.

[k] http://snap.telenet.be/tips/artikel/yelo-tv-app-op-windows-phones-8.

Table 1. Core Living Lab actors for FLELLAP

| Actor | | Role |
|---|---|---|
| Utilisers | FifthPlay | Develop & test city platform |
| | Telenet | Test & develop smart media applications |
| Enablers | IWT | Financial resources |
| Providers | Telenet | Network infrastructure |
| | Alcatel-Lucent | Smart Grid test laboratory |
| | Androme | Integrate solutions in the network |
| | FifthPlay | Smart city platform |
| | iLab.o | Panel management |
| Users | Smart City user panel | Test city platform |
| | General user panel | Profiled panel members |
| Researchers | iMinds | User research |

IWT supports innovation in Flanders in various ways: financial support, services, coordination and policy preparation. First, IWT distributes more than €260 million in subsidies yearly. This money mainly goes to the individual and collective projects of small and large companies, universities, colleges and other Flemish innovation actors. As was already explained in the historical overview of Flemish Living Labs, iMinds and IWT were in charge of redistributing the subsidy that was awarded to the defunct I-City Living Lab. In terms of involvement, the Living Lab manager and the consortium partners had to report to IWT regarding the progress they made once a year. In between, there were also informal contacts between responsibles of IWT and with the Living Lab managers. However, the role of IWT in FLELLAP remained rather passive. In order to stimulate external projects, a one-time call for Living Lab projects was announced within one of the existing funding mechanisms for SMEs of IWT[1], but this resulted in no extra projects for FLELLAP. Interestingly, no other enabler was present in the external cases, as at the time these were conceived, the 'KMO Portefeuille'-option

[1]http://www.iwt.be/evenementen/proeftuinproject-infosessie-door-iwt-ibbt.

(cf. *infra*) was not known, or rather it was not seen as a funding option for Living Lab projects. However, in order to generate research activity and to activate the user panel, the external research cases were largely funded by the researchers themselves who 'invested' their research capacity, thus also acting as a temporary enabler for these cases, and by the utilisers themselves who paid for using the Living Lab infrastructure.

Analysis of the three innovation network paradoxes

- 'dynamic stability' or balancing new relationships and existing relationships.

As the name suggests, this Living Lab network was initially conceived as a platform connecting the separate Living Labs with each their own set of partners, panel and scope. It was intended to attract external projects within these three 'sub-Living Labs' as there were no clearly described internal projects in the project proposal. To this end, the relationships between the actors in the thematic projects need to be strong. Telenet and Androme did build a strong tie within the development of YeloTV, but this remained almost exclusively an exchange of knowledge and (monetary) assets between the two actors. The researchers and panel managers were involved only very late in the innovation process, and the demand came more from their side, so Telenet did not act very committed towards the network. For the Smart Cities Living Lab, clearly more actors were involved (FifthPlay, Telenet, the researchers and the panel managers) and the roll-out of the infrastructure and platform among a dedicated set of end-users was successful, but attracting external utilisers on the FifthPlay platform (new relationships) appeared to be much more difficult. The most interesting and valued use-case on the platform was a smart energy application, which sparked some distrust from FifthPlay towards Alcatel-Lucent, as they were potential competitors in this domain, which was not foreseen in the proposal. However, as Alcatel-Lucent had made a strategic change in priorities, the exploitation of their laboratory infrastructure with the other Living Lab actors remained absent.

Because of these issues, attracting external projects that would lead to new relationships was difficult. This task was mainly executed by the Living Lab Manager and by the iMinds researchers who looked for ways to develop and showcase the research capacity and opportunities, and to activate the panel members. In order to foster a stronger tie with these end-users, the researchers and the panel managers conducted multiple survey waves with fixed time intervals and covering different subjects. This was also carried out in order to gather relevant data to attract external utilisers, but the most important side effect was that the response rates of the panel members remained fairly high, which suggests a sense of community among the end-users. Eventually, three external utilisers engaged in an innovation project (OKEEZ, Fietsnet and MuFoLive), but these projects were executed without any assistance or involvement of the other actors.

## 6. Interaction for Goal Alignment

Because of its intended structure, with three separate Living Lab networks, interaction between the different actors needed to be facilitated in order to reap the potential benefits of the 'platform' structure. This was intended by having regular steering committees to report the progress of the different internal projects, with one Living Lab manager as the central spokesperson. However, in practice this did not work out very well. Because of distrust between some of the partners (FifthPlay and Alcatel-Lucent), partly caused because of the asymmetrical power relationships between the participating actors (SME vs. multi-national) and potential conflicts of interest (Smart Grids & energy), and also because of the rather closed attitude of Telenet and Androme. This resulted in most of the internal as well as external projects being carried out with only a limited set of actors. Only the researchers, panel managers and Living Lab manager had the shared goal to attract external utilisers, carry out innovation projects and activate the user panel, while the companies rather pursued their own goals and agendas.

• Balancing informal and formal relationships
From the different data, we gather that the balance between formal and informal relationships between the different actors within the FLELLAP seemed to be a hurdle. Especially the dual role of Telenet, as utiliser in the smart media thematic domain as well as (potential) provider of its infrastructure in the other cases as well as overall project lead with the Living Lab manager being a consultant who was at that time employed by Telenet. However, because of his position as external consultant working for Telenet, he lacked the negotiation power to mobilise other people and divisions within Telenet for the Living Lab. This was confirmed by the observation that Telenet was not very supportive towards testing its own technologies in FLELLAP as became apparent when discussing the actor roles in the Living Lab (cf. *supra*). Therefore, none of the other companies involved took the initiative to attract external utilisers, but instead they focused on their own projects. As stated earlier, the Living Lab manager, the researchers and the panel managers became an informal alliance in search for external projects, which resulted in three external projects.

In terms of panel members, the FLELLAP panel was rather successful as the response rates of the surveys remained rather high, but the lack of external cases and offline testing activities did not foster a strong sense of community among the panel members.

We will summarise all the previous discussions and analyses by means of the network-related success factors, identified by Rese and Baier[16]: trust, commitment, dependency, and compatibility.

Trust was clearly one of the main issues within FLELLAP, as most of the companies were not eager to share a lot of information regarding their innovations in development. This was also related to the low commitment of some of the actors. Especially the role of Telenet, consortium lead, is notable, as they decided to develop in partnership with another actor from the consortium with the researchers, panel managers and end-users only involved when the innovation was nearly ready to be launched. For Alcatel-Lucent, the commitment went down once the internal company strategy changed. It seems that this problem is mostly related to large companies, as the SMEs were more committed towards their projects, although they also showed less commitment towards the Living Lab itself. This low commitment was also triggered by the fact that none of the actors was heavily dependent on one of the other actors for the projects running in the Living Lab. For the development of the application, Telenet was dependent on Androme, but Telenet decided to take most of the development process outside of the Living Lab. The largest dependency was on the funding by the enabler, but this actor refrained from a lot of participation or control. The (too) broad thematic differentiation also lowered the compatibility of the actors, as they operated in different domains.

## 7. Conclusion

Within our contemporary society, distributed innovation is regarded as the norm rather than as an exception. However, companies still struggle to adequately manage their innovation processes in order to create successful and innovative products and services within this distributed innovation environment. Within this chapter, we have explored Living Labs as open innovation networks that engage in knowledge exchange.

We can conclude that the FLELLAP Living Lab is a clear example of an open innovation network, but that it did not function optimally. As the FLELLAP project proposal did not contain detailed internal innovation projects, and no consortium partners could be categorised as pure utilisers, one of the main goals was to establish a Living Lab infrastructure that would attract external projects and utilisers, and this goal was clearly not reached with only three external projects. Moreover, these projects were not even clearly related to the sub-themes of the Living Lab. One of the major hurdles was the complexity of the constellation because of the different use cases with each of their own dedicated panels. This made it difficult to attract external utilisers, and it also caused that the commitment and dependency amongst the actors was lower, as they had fewer shared goals. It seems advisable to have a more compatible consortium with clearly defined internal projects and a dedicated actor or group of individuals that is concerned with attracting external utilisers to maintain the balance between existing relationships and new relationships. We propose that the enabler(s) of the Living Lab might have more impact on the composition of the network in order to avoid this type of imbalances.

The main limitation of this chapter is that it draws upon the experiences of one Living Lab. Therefore, it is difficult to generalise the insights. However, the complex nature of Living Labs and innovation processes running in these Living Labs lend themselves towards a case study research approach, and this chapter is the first to analyse these processes, set-up, roles and outcomes taking an open innovation network perspective. Future research could also take into account the role of the individuals engaging in this kind of open innovation. This way, the interactions between the different actors within an open innovation network could be studied with the individual as a level of analysis, looking at the impact of individual characteristics on the innovation processes and the outcomes of the network.

## References


1. Chandler, A. D. (1962). *Strategy and Structure* (Vol. 4). Cambridge, MA: MIT press.
2. Lakhani, K. R. and Panetta, J. A. (2007). The principles of distributed innovation. *Innovations*, 2(3): 97–112.
3. Chiaroni, D., Chiesa, V. and Frattini, F. (2011). The open innovation journey: How firms dynamically implement the emerging innovation management paradigm. *Technovation*, 31(1): 34–43.
4. Chesbrough, H. W. (2003). *Open Innovation: The New Imperative for Creating and Profiting from Technology* (Harvard Business Press).
5. Gassmann, O. and Enkel, E. (2004). Towards a theory of Open Innovation: Three core process Archetypes. In *R&D Management Conference*, pp. 1–18.
6. Van de Vrande, V., Lemmens, C. and Vanhaverbeke, W. (2006). Choosing governance modes for external technology sourcing. *R&D Management*, 36(3): 347–363.
7. West, J. and Bogers, M. (2013). Leveraging external sources of innovation: A review of research on open innovation. *Journal of Product Innovation Management*, 31(4) (July 2014), doi: 10.1111/jpim.12125. Available at SSRN: http://ssrn.com/abstract=2195675.

8. March, J. G. (1991). Exploration and exploitation in organizational learning. *Organization Science*, 2(1): 71–87.
9. Van de Vrande, V., De Jong, J. P., Vanhaverbeke, W. and De Rochemont, M. (2009). Open innovation in SMEs: Trends, motives and management challenges. *Technovation*, 29(6): 423–437.
10. Enkel, E., Gassmann, O. and Chesbrough, H. (2009). Open R&D and open innovation: Exploring the phenomenon. *R&D Management*, 39(4): 311–316.
11. Wieczorek, A. J. and Hekkert, M. P. (2012). Systemic instruments for systemic innovation problems: A framework for policy makers and innovation scholars. *Science and Public Policy*, 39(1): 74–87.
12. Smits, R. and Kuhlmann, S. (2004). The rise of systemic instruments in innovation policy. *International Journal of Foresight and Innovation Policy*, 1(1): 4–32.
13. Fichter, K. (2009). Innovation communities: The role of networks of promotors in open innovation. *R&D Management*, 39(4): 357–371.
14. Klerkx, L. and Leeuwis, C. (2009). Establishment and embedding of innovation brokers at different innovation system levels: Insights from the Dutch agricultural sector. *Technological Forecasting and Social Change*, 76(6): 849–860.
15. Fischer, M. M. (2006). The innovation process and network activities of manufacturing firms. *Innovation, Networks, and Knowledge Spillovers: Selected Essays*, 117–133.
16. Rese, A. and Baier, D. (2011). Success factors for innovation management in networks of small and medium enterprises. *R&D Management*, 41(2): 138–155.
17. Dittrich, K. and Duysters, G. (2007). Networking as a means to strategy change: The case of open innovation in mobile telephony. *Journal of Product Innovation Management*, 24(6): 510–521.
18. Westerlund, M. and Leminen, S. (2011). Managing the challenges of becoming an open innovation company: Experiences from Living Labs. *Technology Innovation Management Review* (October 2011).
19. Klerkx, L. and Aarts, N. (2013). The interaction of multiple champions in orchestrating innovation networks: Conflicts and complementarities. *Technovation*, 33(6): 193–210.
20. Dhanaraj, C. and Parkhe, A. (2006). Orchestrating innovation networks. *Academy of Management Review*, 31(3): 659–669.
21. Pittaway, L., Robertson, M., Munir, K., Denyer, D. and Neely, A. (2004). Networking and innovation: A systematic review of the evidence. *International Journal of Management Reviews*, 5(3–4): 137–168.
22. Batterink, M. H., Wubben, E. F., Klerkx, L. and Omta, S. W. F. (2010). Orchestrating innovation networks: The case of innovation brokers in the agri-food sector. *Entrepreneurship and Regional Development*, 22(1): 47–76.
23. Katzy, B., Turgut, E., Holzmann, T. and Sailer, K. (2013). Innovation intermediaries: A process view on open innovation coordination. *Technology Analysis & Strategic Management*, 25(3): 295–309.

24. Almirall, E. and Wareham, J. (2011). Living Labs: Arbiters of mid- and ground-level innovation. *Technology Analysis & Strategic Management*, 23(1): 87–102.
25. Sieg, J. H., Wallin, M. W. and Von Krogh, G. (2010). Managerial challenges in open innovation: A study of innovation intermediation in the chemical industry. *R&D Management*, 40(3): 281–291.
26. Hargadon, A. and Sutton, R. I. (1997). Technology brokering and innovation in a product development firm. *Administrative Science Quarterly*: 716–749.
27. Hargadon, A. and Sutton, R. I. (2000). Building an innovation factory. *Harvard Business Review*, 78(3): 157–166, 217.
28. Verona, G., Prandelli, E. and Sawhney, M. (2006). Innovation and virtual environments: Towards virtual knowledge brokers. *Organization Studies*, 27(6): 765–788.
29. Etzkowitz, H. and Leydesdorff, L. (2000). The dynamics of innovation: From National Systems and "Mode 2" to a Triple Helix of university–industry–government relations. *Research Policy*, 29(2): 109–123.
30. Leydesdorff, L. and Etzkowitz, H. (1996). Emergence of a triple helix of university–industry–government relations. *Science and Public Policy*, 23(5): 279–286.
31. Arnkil, R., Järvensivu, A., Koski, P. and Piirainen, T. (2010). *'Exploring the Quadruple Helix', Report of Quadruple Helix Research for the CLIQ Project, Work Research Centre* (University of Tampere, Tampere, Finland).
32. European Commission. (1995). *Green Paper on Innovation* (European Commission: Brussels).
33. Dosi, G., Llerena, P. and Labini, M. S. (2006). The relationships between science, technologies and their industrial exploitation: An illustration through the myths and realities of the so-called 'European Paradox'. *Research Policy*, 35(10): 1450–1464.
34. Perkmann, M. and Walsh, K. (2007). University–industry relationships and open innovation: Towards a research agenda. *International Journal of Management Reviews*, 9(4): 259–280.
35. Bergvall-Kåreborn, B. and Ståhlbröst, A. (2009). Living Lab: An open and citizen-centric approach for innovation. *International Journal of Innovation and Regional Development*, 1(4): 356–370.
36. Dutilleul, B., Birrer, F. A. and Mensink, W. (2010). Unpacking European living labs: Analysing innovation's social dimensions. *Central European Journal of Public Policy*, 4(1): 60–85.
37. European Commission. (2013a). *Open Innovation 2013* (Luxembourg: Publications Office of the European Union).
38. Frissen, V. and van Lieshout, M. (2004). *To User-Centred Innovation Processes: The Role of Living Labs*. TNO-ICT, Delft.
39. Ballon, P., Pierson, J. and Delaere, S. (2005). Test and Experimentation Platforms for Broadband Innovation: Examining European Practice. In

Conference Proceedings of 16th European Regional Conference by the International Telecommunications Society (ITS), Porto, Portugal, pp. 4–6.

40. Eriksson, M., Niitamo, V. P. and Kulkki, S. (2005). State-of-the-Art in Utilizing Living Labs Approach to User-Centric ICT Innovation — A European Approach. Lulea: Center for Distance-Spanning Technology (Lulea University of Technology Sweden: Lulea).
41. Niitamo, V. P., Kulkki, S., Eriksson, M. and Hribernik, K. A. (2006, June). State-of-the-Art and Good Practice in the Field of Living Labs. In Proceedings of the 12th International Conference on Concurrent Enterprising: Innovative Products and Services through Collaborative Networks (Italy: Milan), pp. 26–28.
42. Kusiak, A. (2007). Innovation: The living laboratory perspective. *Computer-Aided Design and Applications*, 4(6): 863–876.
43. Schaffers, H., Cordoba, M. G., Hongisto, P., Kallai, T., Merz, C. and Van Rensburg, J. (2007). Exploring Business Models for Open Innovation in Rural living labs. In *Proceedings of the 13th International Conference on Concurrent Enterprising. 2007: Sophia-Antipolis, France*, pp. 49–56.
44. Ståhlbröst, A. and Bergvall-Kåreborn, B. (2008). FormIT — An Approach to User Involvement. European Living Labs — A New Approach for Human Centric Regional Innovation, pp. 63–76.
45. Levén, P. and Holmström, J. (2008). Consumer co-creation and the ecology of innovation: A living lab approach. In *Public Systems in the Future: Possibilities, Challenges and Pitfalls*.
46. Feurstein, K., Hesmer, A., Hribernik, K. A., Thoben, K. D. and Schumacher, J. (2008). *Living Labs: A New Development Strategy. European Living Labs — A New Approach for Human Centric Regional Innovation*, pp. 1–14.
47. Turkama, P. (2010). ENoLL Nordic Nordic Network of User-Driven Innovation and Living Labbing Final Report. http://www.nordicinnovation.org/Global/_Publications/Reports/2010/Nordic%20Network%20of%20User-Driven%20Innovation%20and%20Livinglabbing.pdf.
48. Schuurman, D. (2015). Bridging the Gap between Open and User Innovation? Exploring the Value of Living Labs as a Means to Structure User Contribution and Manage Distributed Innovation. Doctoral dissertation. Ghent University. Faculty of Political and Social Sciences; Vrije Universiteit Brussel. Faculty of Economic and Social Sciences; Ghent & Brussels, Belgium.
49. Leminen, S., Westerlund, M. and A.-G. Nyström. (2012). Living Labs as Open-Innovation Networks. *Technology Innovation Management Review*. September 2012: 6–11.
50. Etzkowitz, H. (2008). *Triple Helix Innovation: Industry, University, and Government in Action* (London and New York: Routledge).

51. Cosgrave, E., Arbuthnot, K. and Tryfonas, T. (2013). Living labs, innovation districts and information marketplaces: A systems approach for smart cities. *Procedia Computer Science*, 16: 668–677.
52. Eisenhardt, K. (1989). Building theories from case study research. *Academy of Management Review*, 14(4): 532–550.
53. Yin, R. (1984). Case Study Research. Beverly Hills, CA: Sage Publications.
54. Susman, G. I. and Evered, R. D. (1978). An assessment of the scientific merits of action research. *Administrative Science Quarterly*: 582–603.
55. Ottosson, S. (2003). Participation action research: A key to improved knowledge of management. *Technovation*, 23(2): 87–94.
56. Beyer, H. and Holtzblatt, K. (1999). Contextual design. *Interactions*, 6(1): 32–42.
57. Leminen, S. and Westerlund, M. (2012). Towards innovation in Living Labs networks. *International Journal of Product Development*, 17(1): 43–59.
58. Lievens, B., Baccarne, B., Veeckman, C., Logghe, S. and Schuurman, D. (2014). Drivers for end-users' collaboration in participatory innovation development and living lab processes. In 17th ACM Conference on Computer Supported Cooperative Work (CSCW-2014).
59. Logghe, S., Baccarne, B. and Schuurman, D. (2014a). An exploration of user motivations for participation in living labs. International Society for Professional Innovation Management, Proceedings. Presented at the International Society for Professional Innovation Management (ISPIM-2014).
60. Logghe, S., Baccarne, B., Veeckman, C., Lievens, B. and Schuurman, D. (2014b). Uit passie of voor de poen? Een exploratie van gebruikersmotivaties voor deelname aan innovatie onderzoek in Living Labs. Etmaal van de Communicatiewetenschappen, Proceedings. Presented at the Etmaal 14.
61. Baccarne, B., Logghe, S., Veeckman, C. and Schuurman, D. (2013). Why collaborate in long-term innovation research? An exploration of user motivations in Living Labs. 4th ENoLL Living Lab Summer School 2013, Abstracts. Presented at the 4th ENoLL Living Lab Summer School 2013, Brussels, Belgium: European Network of Living Labs.

# Chapter 10

# OPEN INNOVATION AND TERRITORY

Marcin Baron

*University of Economics in Katowice (Poland)*
*Faculty of Economics*
*Department of Strategic and Regional Studies*

The chapter offers an insight into possible relations between open innovation and territory. While open innovation aims at benefiting of opening the innovation funnel and participating in business ecosystems, territorial (regional) studies and policies stress the role of tangible and intangible resources and ties available in geographical proximity. Both streams, anyway, use a similar toolkit related to knowledge and innovation management. Besides the theoretical framework, some practical examples of open innovation related territorially-bound initiatives are presented in the chapter. Afterwards an outlook on open innovation in territorial policies in Europe is provided.

***Keywords***: Open Innovation; Territory; Territorialisation; Regional Innovation Systems; Smart Specialisations; European Union; Cohesion Policy.

# 1. Introduction

Accessing knowledge of the others and co-creation as well as co-opetition have been always at the very centre of several territorially-bound growth concepts in economics. Be it: "old good" Marshallian industrial districts or its contemporary followers, innovative milieux or clusters. All proving that territory is a prime location for innovation. Therefore the territorial approach to innovation was mainstreamed into policymaking, which was clearly visible in the European Union (EU; formerly European Economic Community) in its structural policy and afterwards in approaches like regional innovation systems or currently promoted smart specialisations. In this context the overall aim of the chapter refers to finding out whether the managerial concept of open innovation (OI) theoretically and practically fits policymaking in the EU. We face paradigm shifts in both regional policy and understanding of OI, but the knowledge landscape and ecosystem issues still may link OI to territory and vice versa. The chapter includes theoretical framework showing possible linkages between OI and territory, examples of territorially anchored initiatives boosting OI and findings of the study concerning the possible presence of OI approaches in the EU cohesion policy 2014–2020.

# 2. Combining OI and Territorial Innovation: Theoretical Frameworks

The widely discussed concept of OI, developed and published by Chesbrough in the early 21st century [2003], has been already mainstreamed in management and economics. Its' author, presenting a number of case studies, proved that some companies radically changed their perception concerning innovation management patterns. Thus, in the modern world, we face a paradigm change from "closed innovation" towards OI. Intensive relationships within business networks, accompanied by multiplicity and diversity of knowledge flows and use of different business models to maximise value creation[1] are key to this approach. The concept of OI in its original assumptions is closely related to the models of R&D activities and focuses on approaching technological change through cooperation with specialised entities in a so called knowledge landscape of a company. Anyway it already faced redefinition done by the authors, who claim that the ecosystem approach becomes core for successful innovation management, with OI thinking being the backbone of decisions upon resource allocation and setting up alliances.

> In our view more interesting strategies emerge when companies combine an open innovation strategy with open business model thinking. A systematic analysis of the possible combinations will lead to an unexpected variety of possibilities, linking "open innovation" and "open business model" strategies to phenomena such as innovation ecosystems.[2]

Moreover, nowadays OI is also often associated with acquiring knowledge from customers and product co-creation. Prosumerism[3,4] and crowdsourcing[5–7] emerge here as iconic buzzwords.

At the same time, the above mentioned assumptions underlying the concept of OI are close to assumptions relevant for the concepts of territorially-bound innovation. In particular, this aspect can be traced in economics and the notion of innovative milieus[8] as well as in theories concerning economic policy exemplified by the concepts of regional innovation systems (RIS)[9] or regional smart specialisations[10] and territory-based ecosystems based upon business ecosystem idea by Moore.[10] Increasing innovation dynamics in territories is inextricably linked to boosting dynamics of R&D cooperation, and recently, following the works of Prahalad and Krishnan[12], also with the inclusion of large social groups in shaping innovative changes.

In a perspective of regional science all issues pinpointed here trace back in time to the last decades of the 20th century. Before this period, the classical location theories had dominated scientific discussion upon the territorial aspect of business. The focus of analysis was primarily on the nature and characteristics of the production process. As such, the essential questions — as noted by McCann and

Sheppard[13] — all had concerned "the ways in which factor inputs were transformed into physical commodities, and this tradition aimed to understand the determining features of this transformation process". Industrial districts by Marshall[14] became an icon of these approaches.

Afterwards the models developed in 19th and early 20th century have been reconsidered. The analysis become more bilateral, challenging the issues of how businesses anchor in territory and how they, in parallel, utilise and enrich their territory. This approach, known as territorialisation[15], emerged upon three phenomena, which have gained significant importance for the economy, i.e., Enabling locality as a response to the challenges of globalisation; overlapping processes of competition and cooperation; understanding human capital and social capital as the main resources in the knowledge economy. Among the other, pinpointed below notions, the "reinvented" industrial districts play significant role here.[16–19]

The concept of innovation became the most important issue within this broad approach. Aydalot and his research group GREMI pioneered the search for "something" that makes it possible for some regions to be more dynamic than the others.[18] The studies primarily concerned: Resources and interactions available to companies in their innovation processes[20]; innovation networks and their spatial aspects[21]; developmental trajectories of regions with the same dominant economic sector.[22] Altogether the works by GREMI, allowed giving the "something" a name of innovation milieu, which is conditioned by three paradigms: The technological paradigm, the organisational paradigm, and the territorial paradigm[18], see Figure 1.

Thus, the logic of this construct says that processes running in territorial proximity benefit of: Being specific, utilising specific know-how, dealing with issues that are better known, mobilising resources in a unique way. If that proves true, businesses well rooted in their milieus should easily enter the OI patterns, especially within their existing networks. In terms of territorial patterns, these kind of networks and spillovers have been widely discussed, especially by Camagni[23] and Capello.[24]

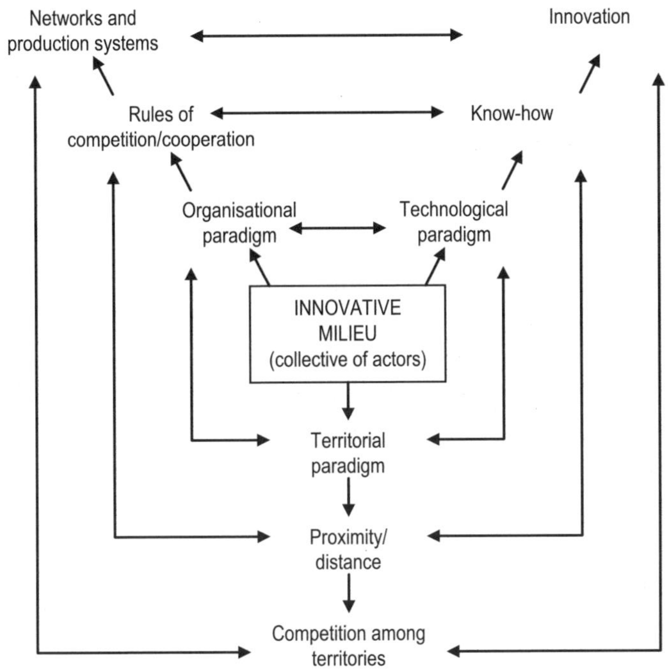


**Figure 1.** The paradigm of innovative milieus and development: Territorialised Economy

*Source*: Ref. [8].

In principles different, but not so far from innovation milieu, is a concept of RIS. When pursuit of innovation was mainstreamed into policymaking, the idea of boosting innovative economy in territories was warmly welcomed. Therefore, the economic notion of innovative milieu was conveyed onto policy level. There it was re-invented as RIS, encompassing innovative companies and their surroundings: partners, competitors, customers, the available human capital, the regional knowledge infrastructure, institutions, regulation and legislation, untraded interdependencies, other factors that influence innovation directly or indirectly as well as external links into the national and global economy.[25] RIS, whatever definition is applied, is perceived as a primary location for innovation.

In the EU, the RIS idea evolved into another Territorial Policy Concept of regional smart specialisations. These specialisations are expected to be place-based innovation policies expressed in dedicated strategies:

> National/regional research and innovation strategies for smart specialisation (RIS3) are integrated, place-based economic transformation agendas that do five important things:
>
> - They focus policy support and investments on key national/regional priorities, challenges and needs for knowledge-based development, including ICT-related measures.
> - They build on each country's/region's strengths, competitive advantages and potential for excellence.
> - They support technological as well as practice-based innovation and aim to stimulate private sector investment.
> - They get stakeholders fully involved and encourage innovation and experimentation.
> - They are evidence-based and include sound monitoring and evaluation systems.[10]

The principles behind achieving regional smart specialisations are summarised in four "Cs", (1) choices and critical mass (focus on limited number of priorities based upon own strengths); (2) competitive advantage (focus on matching research, technology development and innovation capacities with business needs); (3) connectivity and clusters (focus on developing world class clusters that link the region internally and externally); (4) collaborative leadership (focus on efficient innovation systems as a collective endeavour, also giving voice to un-usual suspects).[10]

The word "cluster", used in the above paragraph, itself symbolises a very important notion in research on territorial innovation systems. Dated back to studies on industrial complexes by Czamanski[26,27] it became widely known and followed after the works of Porter[28,29] gained attention not only among the scientific audience. Porter sees clusters as geographic concentration of companies and institutions, mainly in RTD, which are interconnected and work in a common industry. They draw upon a shared pool of talents and skilled, specialised labour as well as utilise specialised infrastructure, services and providers in proximity. Knowledge spillovers boost clusters' dynamics and make it possible for various players to act in a manner of coopetition (i.e., competition blended with cooperation).

In general, all presented approaches pinpoint the role of territory in innovation performance. The nearby located entities are expected to have better conditions for cooperation and enjoy minimised transaction cost, also due to higher level of trust and industry-related, especially knowledge-based, pool of resources. Here a link between regional science and OI theory can be seen. Chesbrough's statement that

> Open innovation is a paradigm that assumes that firms can and should use external ideas as well as internal ideas, and internal and external paths to market, as the firms look to advance their technology.[30]

could so much resemble previous notions if was territorialy-bound. Enkel *et al.*[31] claim that there are three core processes in open innovation: (1) the outside-in process: Enriching the company's own knowledge base through the integration of suppliers, customers, and external knowledge sourcing; (2) the inside-out process: Earning profits by bringing ideas to market, selling IP, and multiplying technology by transferring ideas to the outside environment; and (3) the coupled process: Co-creation with (mainly) complementary partners through alliances, cooperation, and joint ventures during which give and take are crucial for success. And again, if territorial dimension is added to these OI features, no serious differences could be seen compared to the concepts brought in by regional science. Anyway, the both notions are rarely studied jointly.

It is the knowledge landscape that arises as a central issue in a context of linking OI to territory. The managerial approach involves maximising the use of knowledge and its appropriate protection in networks that are not territorially-bound. Today, in the era of globalisation and digitalisation, geographical proximity is not the main criterion for decision making in a perspective of management science, or even in a perspective of microeconomic cost–revenue relationships. Anyway things look different for regional economics. In this case, the essence of activity animation and policy measures is to achieve the maximum accumulation of knowledge within territory. The general objective is therefore related to achieving a situation in which the potential for value creation is sustainably rooted in territory. This approach, of course, evolves over time. A few years ago, it was rather perceived throughout regional R&D partners collaborating with companies to launch innovative products on global market. By this most of the value add could be captured in the territory and build its specialisation. Nowadays, especially after Prahalad and Krishnan's[12] R=G, C=1 (resources are global, customer is individual) concept, focus is also on regional actors' capacity to: Search for key competences in a global scale, capture them, generate the largest share in value add, place advanced products on global markets. Nevertheless, even though the focus may be changed, capturing knowledge and value add remain of primary importance. Having this in mind, in other words, we face a question whether the knowledge landscape of OI-based processes or businesses matches the knowledge landscape catalysed by regional innovation policies.

OI and territorial innovation concepts belong to two different "systems" and as such they are usually studied apart one from another. A short overview upon this is presented in Table 1.

There are few authors like Torkkeli *et al.*[32], Belussi *et al.*[33], Tödtling *et al.*[34] or Amison and Bailey[35] who try to link both dimensions. OI is mainly developed within management/innovation management and is perceived principally as bigger player's approach (strategy) to achieving knowledge, ideas, components from inventors and innovators. This approach is characteristic for Chesbrough and numerous followers. As a consequence, these authors hardly ever link OI to territory. On the other hand, accessing knowledge of others and co-creation as well as co-opetition have been always at the very center of territorially anchored: industrial districts, innovative milieus or clusters. All the mentioned concepts got from Isard's regional science into (regional) policymaking. And due to this fact were generously offered various public support schemes

**Table 1.** Open innovation and territorial innovation — key characteristics

| | Open innovation | Territorial innovation |
|---|---|---|
| Scientific background | Management, innovation management | Economics, regional science |
| Orientation | • Business model perspective<br>• Knowledge landscape<br>• Effectiveness and profitability | • Territorial perspective<br>• Knowledge spillovers<br>• Competitiveness and reduction of transaction costs |
| Objective | Benefiting of opening the innovation funnel and/or of participating in business ecosystems | Benefiting of pool of tangible and intangible resources and ties available in geographical proximity |
| Notions | • Open innovation<br>• Open innovation 2.0<br>• Open business model<br>• Innovation ecosystem | • Industrial districts<br>• Innovation milieu<br>• Clusters<br>• Regional innovation systems<br>• Regional smart specialisations<br>• Regional innovation-oriented ecosystems |

that use toolkits quite similar to OI. In other words, we face two important scientific streams proposing similar tools to deal with knowledge, ideas and resources transfer aimed at innovativeness. And this streams usually not merge. The key difference is that the OI stream is biased towards in-house innovation management strategy and tactics, while the territorial innovation stream is biased to networking as a co-ordination mechanism. Consequently, the first one refers to business model and corporate aims, while the latter addresses co-operation and aims of the network and its territory. Tackling both of them — as partly shown by Torkkeli *et al.*[32] — is worthy, especially though for many start ups their territorial milieus are of a first choice to raise capital by entering the first OI patterns. This perspective — of smaller players and their strategies to utilise (or not utilise) OI possibilities in the nearest area — is therefore fairly unique and surely interesting in terms of both academic research and practice.

## 3. Placing OI in Territory

Before some practical examples of OI-related territorially-bound initiatives are shown, it is worth to pinpoint a general observation done by Amison and Bailey.[35] They call attention to the idea of Phoenix industries understood as clusters of small- and medium-sized businesses, working with broadly similar technologies, that have sprung up in former industrial areas. The authors claim that these territories, due to historic, relatively immobile, investments in industry knowledge and workforce skills became resourceful for businesses which research, develop and produce sophisticated components for use in a range of industries. In such environments, they see huge potential for OI. Again, this approach shows that findings presented in Section 2 are relevant.

Phoenix industries were cited here on purpose as a list of exemplary OI initiatives placed in territories will be started with UK's "Niche Vehicle Network", scrutinised by Amison and Bailey.[35] Practically, the idea behind the network is to enable potentials based on old automotive and related manufacturers in the Midlands — where 700,000 manufacturing jobs have disappeared from the region since the 1960s. There are two driving forces for the concept. The first one is connected to Jaguar and Land Rover operations. The latter refers to automotive suppliers of various sectors who internationalise by offering their quality products to the other markets, e.g., German or Japanese.

> The Niche Vehicle Network is an independent association of over 100 niche vehicle manufacturers, specialist technology companies and supply chain. Established in 2005, the Network is free to join and all events free to attend. It promotes the development and application of new technology, by bringing together independent vehicle manufacturers, system suppliers, automotive technology companies and HEIs, to collaborate on the innova-

tive application of technologies in low-volume vehicle production. [www.nichevehiclenetwork.co.uk]

According to the programme website, 104 different business collaborations have been catalysed and the network has assisted the development of 11 prototype vehicles and 3 process improvement projects. So called "Collaborative Research and Development Projects" are funded by the network. The exemplary focus areas are: Lightweight vehicles, range extending of electric vehicles, fuel cell hybrid applications, hydrogen and electric taxis, zero emission sports cars, low investment chassis assembly processes. Aerodynamic studies and design guidance are also provided within the network.

Thematically similar is a network of automotive producers in South Poland. It is founded by Katowice Special Economic Zone (governmental company established to offer investment plots and tax incentives in several locations around old industrial city of Katowice) and two consulting companies. Automotive sector is the most important player in the zone. For this reason its board established "Silesia Automotive" as an anchorage instrument. The initiative is based on two pillars: (1) Human resources and labour market and (2) new materials and engineering. The second pillar utilises OI toolkit to focus, for example, on: materials joining, composite and composite shells, electric and electronic applications, prototyping. Process issues like: Product development, energy efficiency, lean manufacturing, etc., are also covered by the initiative.

Another OI initiative that builds upon industrial heritage of the territory is "Chemelot" in Sittard-Geleen, Limburg (Netherlands). There the coal mine that existed in 1926–1967 catalysed the creation of the coking plant which further developed into a chemical company DSM operating since the 1930s. A range of products were offered by DSM throughout the 20th century. Year 2000 marked the beginning of a new era in which the company was led away from bulk chemicals and polymers in the direction of specialty products in performance materials, health and nutrition. In the meantime, petrochemical units were sold to an external investor from Saudi Arabia and another divestments followed. Altogether all these processes ended up with creation of "Chemelot" industrial park (utilising the existing industrial infrastructures) as well as campus and venture capital fund strongly oriented on OI in chemicals.

> Chemelot Campus accommodates approximately 1200 knowledge workers who are engaged in the development of new products, product applications and production processes and in chemistry education. Since 2005, dozens of new companies active in research, development and production of CHEMaterials have opened units at Chemelot, attracted by the expertise, facilities, utilities and permits available there. In addition, dozens of new service providers have set up shop there to support new and existing companies in their activities. [www.chemelot.nl]

In case of "Chemelot" the links to R&D and education centers, especially in nearby Maastricht, Aachen and Leuven are of utmost importance. The management promotes the idea of open and continuous innovation concerning: technology, products and processes as well as referring to business models and organisational structures.

Anyway, OI-related territorial activities are obviously not only focused on industrial redevelopment. In Swedish Lund (Scania), the prominent businesses and institutions: Ericsson, TeliaSonera, Sony Mobile, Lund University, Malmö University, Blekinge University and Regional Council of Scania established an organisation named "Mobile Heights". This member financed institution is open for: companies, industry associations, academic institutions and public organisation. Achieving excellence in mobile innovation is the key idea. The partnership offer, apart of the networking, includes research programmes and multi-disciplinary innovation projects. They are supported by trend analysis and expertise. A business accelerator "Mobile Heights Business Center" since 2009 adds services for start-ups and organises events for the business community. Another step is being done with recent opening of the "Mobile Heights Center" building, which is going to be:

> a modern and open venue to promote collaboration between companies and organisations in the southern region — already today

> showcasing a mix of established industry to the newest innovation companies and start-ups and in addition to academic institutes and organisations within the innovation system. [...] Mobile Heights Center offers a unique, innovative environment. We've been developing mobile communication here already for many years. It's no coincidence that Mobile Heights Center is situated in Skåne, in the city of Lund. That's because here there is access to unrivaled competence representing the entire value chain — from hardware and antennae manufacturers to software developers. [www.mobileheightscenter.com]

Even though the concept was marked as not being the industrial redevelopment, it must be anyway admitted that it becomes somehow a way of tackling structural problems in regional economy, with use of OI approach. It is so, because the Center was created in the former premises of Sony Mobile. According to the press release [http://evertiq.com/news/22725] in 2012, this company redefined its global strategy which affected Lund with approximately 650 jobs closures. Anyway the location was still planned to be an important strategic site for Sony Mobile, with the main focus on software and application development. About 1,000 workplaces are expected to emerge thanks to the renewed facility.

OI initiatives may also link neighbouring territories as it is being done in an "Open Alps" project co-funded by the EU within the Alpin Space Programme. The initiators of the project aim at fostering the spread of open innovation practices among SMEs located in the Alpine area of Italy, Austria, France, Germany and Slovenia. A database of SMEs and R&D institutions was created to accompany an OI web-based platform for innovation seekers and providers:

> As a so-called "innovation seeker" you can launch calls and publish your innovation requests. The degree of information disclosure as well as the advertised reward for successful solutions is completely up to you — but of course we are happy to give you our support and advice!

> In case you are a small or medium-sized enterprise or a research and development centre [...] you can use our platform to present your expertise. As a first step you can take a closer look at the calls for innovation solutions. Secondly, you can pick the calls that raise your interest, elaborate and submit your own solution proposals. In case the issuing company is intrigued by your offer we will bring you in touch and assist you with starting joint activities. [www. alpine-open-innovation.eu]

"Open Alps" is especially focused on social and eco-innovation. Apart of the platform it offers an array of territorially-bound liaison activities, i.e., local OI support centres to facilitate the access to R&D institutions and strengthen networks; local OI labs consisting of SMEs and research and development centres; OI management trainings.

Finally, the reader's attention may be attracted by a specific regional–global university-based OI initiative named "Demola". It aims at building the World's strongest innovation ecosystem by linking regional OI hubs at universities. The concept is illustrated in Figure 2.

| **Demola is:** | |
|---|---|
| **a network** that consists of various partners including universities, their faculties, researchers and students, as well as companies, local agencies and a growing number of Demola Centers around the globe. Not only are we international, we are interdisciplinary. | **a co-creation** concept that is geared to solve real challenges. Every project has an outcome – be it a new concept, a demo, or a prototype. If the partner company finds the outcome useful, the company can license or purchase the outcome, and take it for further development. |
| **a process** that is formatted and facilitated. The Demola process ensures that the work is systematic and runs on schedule. This way, the work itself can be as creative as possible, but the process keeps things under control both in terms of time and deliverable. | **a framework** that makes it easy for partners to come in and cooperate. Each partner has a clear role, and the work is guided by simple procedures. Contracts, intellectual property rights, licensing models, and other legal requirements are in place and meet international business standards and practices. |

**Figure 2.** The "Demola" concept
*Source*: www.demola.net.

As it is reported for 2014, "Demola" centres in 9 locations linked 140 partner companies, 1,600 students and 37 universities in 350 projects.

The presented cases show that OI-based initiatives are being placed in various territorial contexts: From regions fighting their path dependency to larger areas boosting internal cooperation or local business ecosystems entering global networks. It must be, anyway, honestly pinpointed that the OI notion is not itself the sole driving force for the presented concepts. Several tools, nowadays branded as OI, had been utilised in territorial economic development long years before Chesbrough's ideas were published.

## 4. OI in Territorial Policies in Europe

The message of the preceding paragraph in theoretical or scientific classification terms blurs the scene. But this confusion is inevitable. The reason for that has been already given. OI as a concept is relatively new, compared to the presented territorial concepts and their related policy approaches. Therefore, for obvious reasons, some of the existing innovation management techniques and tools were incorporated into OI thinking and the other way, some of the existing territorial initiatives or toolkits received the fancy OI label. So the question emerges, whether OI is or can be a subject of territorial, mainly regional, policy.

The study by Tödtling *et al.*[34] shows that (1) there is no uniform model of open innovation that applies to all types of regions, (2) region-specific solutions and policies are required that account for divergent regional conditions, and (3) change towards a culture facilitating OI takes considerable time. The author's own study[36] of IT companies in selected Polish regional innovation systems revealed that businesses are rather reluctant in utilising OI-related opportunities. Anyway, as it is concluded:

> in case of entities that open they innovation patterns, territorial range of activities is not of primary importance. More interactions are visible on regional level, compared to extra-regional, which is

> not surprising for SMEs. But these interactions, linked to presence of OI processes in business routines, are rather not in favour of any (regional, extra-regional) options.[36]

According to the study, the regional dimension plays an important role in case of idea or technology shared during meetings; whereas licences are mostly acquired extra-regionally. This may origin from both regional characteristics and global specifics of enabling technologies in IT.

In order to propose a policy mix for OI in territorial perspective the general philosophy of OI can be cross-fertilised with Laranja's[37] functional approach to regional innovation systems. By doing so, a following list of policy actions targeted at businesses as well as research and development sector can be extracted:

- Participation in conferences and seminars;
- participation in professional training and HR programmes;
- cooperation with universities in education or research;
- participation in associations or clustering initiatives;
- joint actions or strategies in the industry;
- participation in: ideas exchange, troubleshooting, companies presentation meetings;
- using online: ideas exchange, troubleshooting, companies presentation tools;
- using technology brokerage/transfer services.

All these actions are willingly used by regional policymakers or business support organisations in their toolkits. In the EU, they have been generously co-financed with cohesion policy funds (European Regional Development Fund [ERDF], European Social Fund [ESF]) as well as under the research and innovativeness agenda (Framework Programmes). Especially in the two programming periods influenced by the (unfortunately unsuccessful) Lisbon Strategy for 2000–2010, i.e., 2000–2006 and 2007–2013, the EU contributed a lot to establishing relevant initiatives across the European territory. Starting from the most disadvantaged regions to the highly competitive areas, which were supported to strengthen their global position.

Boosting innovation remains also in the current budget perspective of the EU — encompassing years 2014–2020. The following paragraphs are expected to answer a question, whether OI will be/may be specifically enabled within. In general innovation is mainstreamed in the EU's policy as, together with knowledge, it is a cornerstone of so called smart growth. The existing strategy of the EU — Europe 2020 — puts forward three mutually reinforcing priorities: smart growth, sustainable growth and inclusive growth. The smart component contains a policy flagship initative "Innovation Union". Under this initiative, the European Commission is obliged, among the others, to:

> improve framework conditions for business to innovate (i.e., create the single EU Patent and a specialised Patent Court, modernise the framework of copyright and trademarks, improve access of SMEs to Intellectual Property Protection, speed up setting of interoperable standards; improve access to capital and make full use of demand side policies, e.g., through public procurement and smart regulation);
>
> promote knowledge partnerships and strengthen links between education, business, research and innovation, including through the EIT, and to promote entrepreneurship by supporting Young Innovative Companies; [Europe 2020, 2010]

which are rather general commitments showing not more than the possibility for OI notions to be easily included in or easily excluded from the political agenda.

Anyway, in terms of territorial policies the key enabling role at the EU level is played by a bundle of regulations concerning the resource allocation, setting frameworks for ERDF and ESF. The three regulations of the European Parliament and of the Council [No. 1303/2013; No. 1301/2013; No. 1304/2013 all of 17 December 2013] operate with a logic of thematic objectives decomposed into investment priorities. Those of them, which could be considered important for increasing the dynamics of OI, have been listed in Table 2.

**Table 2.** EU's Cohesion Policy 2014–2020 investment priorities relevant for open innovation

| Thematic objective | Investment priority and fund |
|---|---|
| Strengthening research, technological development and innovation | • Enhancing research and innovation (R&I) infrastructure and capacities to develop R&I excellence, and promoting centers of competence, in particular those of European interest; ERDF<br>• Promoting business investment in R&I, developing links and synergies between enterprises, research and development centers and the higher education sector, in particular promoting investment in product and service development, technology transfer, social innovation, eco-innovation, public service applications, demand stimulation, networking, clusters and open innovation through smart specialisation, and supporting technological and applied research, pilot lines, early product validation actions, advanced manufacturing capabilities and first production, in particular in key enabling technologies and diffusion of general purpose technologies; ERDF |
| Enhancing access to, and use and quality of, ICT | • Extending broadband deployment and the roll-out of high-speed networks and supporting the adoption of emerging technologies and networks for the digital economy; ERDF |
| Enhancing the competitiveness of SMEs | • Promoting entrepreneurship, in particular by facilitating the economic exploitation of new ideas and fostering the creation of new firms, including through business incubators; ERDF<br>• Developing and implementing new business models for SMEs, in particular with regard to internationalisation; ERDF<br>• Supporting the creation and the extension of advanced capacities for product and service development; ERDF<br>• Supporting the capacity of SMEs to grow in regional, national and international markets, and to engage in innovation processes; ERDF |

(*Continued*)

Table 2. (*Continued*)

| Thematic objective | Investment priority and fund |
|---|---|
| Supporting the shift towards a low-carbon economy in all sectors | • Promoting research and innovation in, and adoption of, low-carbon technologies; ERDF |
| Preserving and protecting the environment and promoting resource efficiency | • Promoting innovative technologies to improve environmental protection and resource efficiency in the waste sector, water sector and with regard to soil, or to reduce air pollution; ERDF<br>• Supporting industrial transition towards a resource-efficient economy, promoting green growth, eco-innovation and environmental performance management in the public and private sectors; ERDF |
| Promoting sustainable and quality employment and supporting labour mobility | • Supporting the development of business incubators and investment support for self-employment, micro-enterprises and business creation; ERDF<br>• Self-employment, entrepreneurship and business creation including innovative micro, small- and medium-sized enterprises; ESF |
| Investing in education, training and vocational training for skills and lifelong learning | • Enhancing equal access to lifelong learning for all age groups in formal, non-formal and informal settings, upgrading the knowledge, skills and competences of the workforce, and promoting flexible learning pathways including through career guidance and validation of acquired competences; ESF |

*Sources*: Compiled of Regulations of the European Parliament and of the Council No. 1303/2013, No. 1301/2013, No. 1304/2013 of 17 December 2013.

Two general observations can be made with reference to the findings above. The first pinpoints the leading role of ERDF financing in processes potentially enabling OI. The other calls our attention to the fact that even though OI is literally linked to the thematic objective "Strengthening research, technological development and innovation", there are several other headlines where OI patterns can be easily applied.

It must be also noted that in principle OI, in policy terms, will be supported if related to smart specialisations. For the reader, this recommendation creates a loop to Section 2 in which the concepts of OI and territorial innovation are being bound. Here territorially-bound OI is understood as one of the policy instruments.

Another important step in policy programming will be taken with final approval of partnership agreements between the European Commission and individual EU countries. These documents set out the member states' plans on how to use funding from the European budget between 2014 and 2020. They outline each country's strategic goals and investment priorities and present the scope of operational programmes elaborated to structure allocation. The final shape of operational programmes in countries and regions, when approved by the European Commission — late 2014 or early 2015, will reveal the scope of actions enabled in specific territories. Some general directions, for the moment the book is being edited, can be outlined with use of the position papers, which were presented to every member state as the Commission's views in 2012. The key recommendations in perspective of potential OI applications have been summarised in Table 3.

The statements of general nature, like for example:

> Given the country's weaknesses in this particular area, funds should concentrate their interventions promoting business R&D and innovation investment, product and service development, technology transfer, networking, clusters and open innovation through smart specialization. Investments in research and innovation should follow a multi-level and business-led Research and Innovation Strategy for Smart Specialization. [Position Paper for Portugal]

have been purposely not included in Table 3. It is because its objective was to present some focus areas that are really, in strategic terms, open certain niches concerning expected innovation advances in the country and its regions. For this unique ideas, OI toolkit can be easily applied to territorial development. The plethora of other

**Table 3.** European Commission's recommendations concerning innovation policies in the member states in perspective of potential open innovation application

| Country | Selected recommendations |
|---|---|
| Austria | • Austria should enhance business innovation and competitiveness in agriculture, forestry and aquaculture. It is essential to improve agricultural productivity through research, knowledge transfer, cooperation and innovation.<br>• Austria should invest more in sustainable energy technologies and climate research, with a view to promote a resource-efficient and low carbon-economy. This requires the promotion of innovation in eco-technologies and renewable energy technologies, of energy efficiency in all sectors in order to reduce $CO_2$ emissions, and of the decarbonisation of urban and rural transport as well as supporting innovative traffic management policies. |
| Belgium | • Given the challenges it faces in terms of both cost- and non-cost competitiveness, Belgium should pursue a two-pronged strategy to boost its overall economic performance. Increased investment in key enabling technologies, research and development (R&D), entrepreneurship and marketable innovation would strengthen Belgium's position in the world's good and services markets by increasing the added value of its output. |
| Bulgaria | *No specific recommendations found* |
| Croatia | • Enterprises should be supported in the start-up stage and during their lifecycle aiming at an increased productivity and competitiveness, development of new products and processes, up-taking of ICT and achieving more efficiency in terms of energy and resource consumption involving innovative and greener production processes. |
| Cyprus | • Enhancing competitiveness and innovation of SMEs in all sectors (including agriculture and fisheries) should be a priority. Support should become more targeted, focussing on business innovation and the use of new eco-efficient technologies, including linkages to good practice and enterprises in other countries. |

(*Continued*)

**Table 3.** (*Continued*)

| Country | Selected recommendations |
|---|---|
| Czech Republic | *No specific recommendations found* |
| Denmark | • In order to ensure the European added value, Denmark should identify the investments in these areas, which would not be financed otherwise. An example is eco-innovation, which constitutes an important development opportunity for Denmark.<br>• Denmark has had good experience in fostering clusters and networking in both B2B (Business to Business) and in a triple-helix context. The funds should continue to support these efforts via e.g., technology transfer, social innovation and public service application, demand stimulation, networking, clusters and open innovation through smart specialisation. Especially on cluster-building and cooperation, the wider, cross-border cooperative structures like the Baltic Sea Strategy, offer possibilities both to use the experience, which foreign actors have in operating on non-Danish markets, and to bring knowledge and innovation to Denmark in a strategic way. |
| Estonia | *No specific recommendations found* |
| Finland | • The main elements of broadening the innovation base relate to the ongoing structural change of the Finnish economy and export sectors. The change offers renewal opportunities and growth potential to the R&I intensive ICT sector. Investments should support wider participation of all businesses in regular innovative activities, also on sparsely populated areas.<br>• Technological and applied research, pilot lines, early product validation actions, advanced manufacturing capabilities and first production in Key Enabling Technologies and diffusion of general purpose technologies provide opportunities for Finnish SMEs to utilise the extensive R&I know-how of the country.<br>• Clean technology and eco-innovations provides good opportunities for broadening the Finnish economic base and commercialisation of R&I inputs. Developing commercialisation requires specific attention to be given to stimulating innovation in SMEs and increasing private sector involvement in the design and implementation of actions. |

(*Continued*)

Table 3. (*Continued*)

| Country | Selected recommendations |
|---|---|
| Germany | • The priority in research & innovation area should be the creation of an innovation-friendly business environment in order to support the shift towards an innovation-driven economy in particular to address regional disparities. The funds should be allocated according to regional and specific needs. The potential for growth in the maritime economic sectors should be further explored, as identified in the Blue Growth initiative.<br>• The funds should invest in eco-innovation, new types of business organisations and greening the economy and resource and energy efficiency, e.g., by putting in place incentives that stimulate companies to measure, benchmark and improve their resource efficiency systematically. It should be ensured that advice and support is available to help SMEs to exploit their potentials for energy savings and efficiency and sustainable use of raw materials. |
| Greece | *No specific recommendations found* |
| Hungary | • A specific research strategy for the environment and agricultural sectors is needed to contribute to growth and job creation by improving research and innovation related to climate change actions, eco-innovation and sustainable use of resources. Funds should also promote increased cooperation between rural SMEs/ the agro-industrial sector on the one hand, and specialised research centers on the other so as to stimulate knowledge spill-overs, technology transfer and innovation. |
| Ireland | • Funds should encourage the promotion of cooperation with a view to creat and develop clusters especially Regional Maritime Clusters as drivers of Blue Growth and for the transfer of marine knowledge and technology to industry and supported, where necessary, in a territorial cooperation context.<br>• Funds should assist the development and, where possible, the deployment of innovative technologies and equipment leading to the production of renewable energy, including marine, wind and biomass energies.<br>• As regards the distribution of renewable energies, the development and deployment of intelligent metering systems and smart grids should be fostered, with the corresponding provision of training in order to meet the skill development needs in this area. |

(*Continued*)

Table 3. (*Continued*)

| Country | Selected recommendations |
|---|---|
| Italy | • Enhancing innovation is required in traditional industrial sectors in order to maintain competitiveness. Coordination and joint initiatives in this sector should be carried out with the neighbouring countries and/or regions. Key industries (manufacturing, mechanical, motor vehicles, agro-industry, transport, office machinery and computers and equipment, media and communication) as well as traditional sectors (culture, tourism, agriculture, textile and clothing, leather and footwear, furniture and wood products) need to boost their productivity and exports in order to avoid further loss of national and international market shares.<br>• With regard to the competitiveness of rural SMEs and the agricultural and agro-food sector, priority should be given to increase added value and productivity through innovation. Innovative relations between primary agricultural producers and other economic actors, also from traditionally unrelated economic areas, should be promoted, while knowledge and innovation transfer should be a priority for less advanced regions. |
| Latvia | *No specific recommendations found* |
| Lithuania | *No specific recommendations found* |
| Luxembourg | *No specific recommendations found* |
| Malta | • Interventions should focus on enhancing research and innovation (R&I) infrastructure to develop R&I excellence and promoting centers of competence in particular those of European interest (such as a Centre of Excellence on Adaptation to Climate Change or Energy Efficiency and Green Technologies).<br>• The indigenous private sector R&I needs to be stimulated by concentrating resources and specialising within key sectors, identifying niche areas linked to existing/potential entrepreneurial activities on which to build critical mass. Sectors that have already been identified include health and biotechnology, energy, climate and environmental technologies, ICT and high value-added manufacturing, aviation, businesses based on intangible assets such as cultural and creative industries as well as the maritime and aquaculture sectors. |

(*Continued*)

**Table 3.** (*Continued*)

| Country | Selected recommendations |
|---|---|
| The Netherlands | • Funds should mainly induce private R&I investment needed by SMEs and growing companies to develop or improve their in-house research facilities, know-how and skills. |
| Poland | *No specific recommendations found* |
| Portugal | *No specific recommendations found* |
| Romania | *No specific recommendations found* |
| Slovakia | *No specific recommendations found* |
| Slovenia | *No specific recommendations found* |
| Spain | *No specific recommendations found* |
| Sweden | *No specific recommendations found* |
| United Kingdom | • The Funds should support innovative production and consumption patterns that minimise the negative impact on the environment, particularly among SMEs, including in the agriculture, fisheries and aquaculture sectors, and promote innovative ways of using ecosystem-based approaches to improve the competitiveness of SMEs. |

*Source*: Compiled of Position Papers of the Commission Services on the development of Partnership Agreement and programmes in the member states.

suggestions, also for the countries marked with "No specific recommendations found" were not quoted. Even though many known tools like: Business incubation, supporting internationalisation, funded research and development programmes etc. could also be OI-labelled if set up in a specific way.

This policy overview allows also further reflections upon the possible readiness level towards territorially-based OI applications. Assuming that both member states' and Commission's expectations meet in the Position Papers, some observations can be done. Firstly, in some of the countries (e.g., Bulgaria, Croatia, Romania, Spain), the focus of political intervention in territories is on social issues or infrastructure, not on innovation. These countries usually have some general approaches to innovation issues. The other countries (e.g., Latvia, Lithuania, Poland, Slovakia) still catch-up and try to fix as many issues as possible with the use of external money even though their plans are rather of a general nature. There is also a group of countries that (still?) focus on setting up a system, relevant public administration patterns etc. (Greece is a leading example here), believing that sound governance will boost innovativeness and competitiveness. Finally, there is a group of experienced players, who mostly get limited EU cohesion policy funding due to their overall high economic performance. In these countries (e.g., Germany, the Netherlands, Austria, Belgium), maybe due to smaller sums available, approaches are much more focused and some of the recommendations sound as ready-to-use themes for regional or national OI actions.

## 5. Closing Remarks

OI and territorial innovation origin from different fields of studies and research interests. While OI aims at opening the innovation funnel and/or participating in business ecosystems, the latter notion is about benefiting of pool of tangible and intangible resources and ties available in geographical proximity. Knowledge — its landscape and spillovers — is a cornerstone allowing joint understanding of both approaches: seeing OI in a territory. Obviously, it is not a must that OI must be placed in territory (usually it is not), neither territorial innovation must utilise OI. In numerous cases, OI can constitute a toolkit for boosting innovative dynamics of territories. It is the policy or business decision whether to apply OI or choose different solutions. On the other hand, it is the corporate/managerial decision concerning the level of territorial anchorage of the OI processes, to be undertaken by companies willing to follow the OI pathway. If OI and territory are linked together another link emerges in most of the cases — it is a link to policy. In the EU a sound support is given to national and regional innovation policies, anyway — first of all — this support is provided to different pro-innovative approaches, not only OI; besides not all EU member states seem to have a clear understanding of their political agenda and action plans concerning the innovation policy. The question remains, whether the old member states seem to be better focused, because: (1) limited external resources are available for them, (2) specialisation is a game for the rich or (3) it takes really a long time to build capacity for targeted innovation policy?

## Acknowledgement

The first written reflections on this subject were prepared for The XXV ISPIM Conference. To introduce the empirical study on Polish SMEs, parts of this text has been already delivered to the conference proceedings [Baron, M. (2014). Open Innovation cooperation strategies in Regional Innovation System. *Proceedings of The XXV ISPIM Conference — Innovation for Sustainable Economy & Society*, Dublin, Ireland on 8–11 June 2014].

## References

1. Chesbrough, H. (2006). *Open Business Models: How To Thrive In The New Innovation Landscape*, 1st Ed. (Harvard Business Review Press).
2. Vanhaverbeke, W. and Chesbrough, H. (2014). *New Frontiers in Open Innovation*, Chesbrough, H., Vanhaverbeke, W. and West, J. (eds.) Chapter 3 "A Classification of Open Innovation and Open Business Models" (Oxford University Press).
3. Jacobfeuerborn, B. (2014). An Informational model of open innovation; intelligent tools for building a scientific information platform: From research to implementation. *Studies in Computational Intelligence*, 541: 15–24.
4. Di Gangi, P. M. and Wasko M. (2009). Open innovation through online communities, knowledge management and organizational learning. *Annals of Information Systems*, 4: 199–213.
5. Sloane, P. (ed.) (2011). *A Guide to Open Innovation and Crowdsourcing: Practical Tips Advice and Examples from Leading Experts in the Field* (Kogan Page Limited).
6. Leimeister, J. M., Huber, M., Bretschneider, M. and Krcmar, H. (2009). Leveraging crowdsourcing: Activation-supporting components for IT-based ideas competition. *Journal of Management Information Systems*, 26(1) Summer: 197–224.
7. Brunswicker, S. and van de Vrande, V. (2014). *New Frontiers in Open Innovation*, Chesbrough, H., Vanhaverbeke, W. and West, J. (eds.) Chapter 7 "Exploring Open Innovation in Small and Medium-Sized Enterprises" (Oxford University Press).

8. Crevoisier, O. (2004). The innovative milieus approach: Toward a territorialized understanding of the economy? *Economic Geography*, 80(4): 367–379.
9. Braczyk, H.-J., Cooke, P. N. and Heidenreich, M. (1998). *Regional Innovation Systems. The Role of Governance in a Globalized World* (UCL Press, London).
10. Foray, D., Goddard, J., Goenaga Beldarrain, X., Landabaso, M., McCann, P., Morgan, K., Nauwelaers, C. and Ortega-Argilés, R. (2012). *Guide to Research and Innovation Strategies for Smart Specialisation (RIS 3)*, European Union, Smart Specialisation Platform.
11. Moore, J. F. (1993). Predators and prey. A new ecology of competition. *Harvard Business Review*, May–June.
12. Prahalad, C. K. and Krishnan, M. S. (2008). *The New Age of Innovation: Driving Cocreated Value Through Global Networks*, 1st Ed. (McGraw-Hill).
13. McCann, P. and Sheppard, S. (2003). The rise, fall and rise again of industrial location theory. *Regional Studies*, 37(6–7): 649–663.
14. Marshall, A. (1890). *Principles of Economics* (Macmillan, London).
15. Jewtuchowicz, A. (2005). *Terytorium i współczesne dylematy jego rozwoju* (Wydawnictwo Uniwersytetu Łódzkiego, Łódź).
16. Becattini, G., Bellandi, M., Dei Ottati, G. and Sforzi F. (2003). *From Industrial Districts to Local Development. An Itnerary of Research* (Edward Elgar, Cheltenham UK, Northampton USA).
17. Dei Ottati, G. (2002). Social concentration and local development: The case of industrial districts. *European Planning Studies*, 10(4): 449–466.
18. Bellandi, M. (2002). Italian industrial districts: An industrial economics interpretation. *European Planning Studies*, 10(4): 425–437.
19. Brusco, S. (1992). *Industrial Districts and Local Economic Regeneration*. Pyke, F. and Sengenberger, W. (eds.) Chapter "Small Firms and the Provision of Real Services" (International Institute for Labour Studies, Geneva).
20. Maillat, D. and Perrin, J.-C. (eds.) (1992). *Enterprises Innovatrices et Développement Territorial* (GREMI EDES, Neuchâtel).
21. Maillat, D., Quévit, M. and Senn, L. (eds.) (1993). *Réseaux d'innovation et milieux innovateurs: Un pari pour le développement regional* (GREMI EDES, Neuchâtel).
22. Ratti, R., Bramanti, A. and Gordon, R. (eds.) (1997). *The Dynamics of Innovative Regions. The GREMI Approach* (Aldershot, Ashgate).
23. Camagni, R. (1991). *Innovation Networks: Spatial Perspectives*, Camagni, R. (ed.) "Introduction: From the Local 'Milieu' to Innovation Through Cooperation Networks" (Belhaven Press, London).
24. Capello, R. (1999). Spatial transfer of knowledge in high technology milieux: Learning versus collective learning processes. *Regional Studies*, 33(4): 353–365.
25. Martin, R. L. (2003). *A Study on the Factors of Regional Competitiveness* (Report for The European Commission Directorate-General Regional Policy, Cambridge Econometrics).

26. Czamanski, S. (1974). *Study of Clustering of Industries*, with the assistance of Czamanski, D. Z. and Ellis S. B. (Institute of Public Affairs, Dalhousie University, Halifax).
27. Czamanski, S. (1976). *Study of Spatial Industrial Complexes* (Institute of Public Affairs, Dalhousie University, Halifax).
28. Porter, M. (1998). Clusters and the new economics of competition. *Harvard Business Review*, 76 (November–December): 77–90.
29. Porter, M. (1990). *The Competitive Advantage of Nations* (Free Press, New York).
30. Chesbrough, H. (2003). *Open Innovation. The New Imperative for Creating and Profiting from Technology* (Harvard Business School Press, Boston, Massachusetts).
31. Enkel, E., Gassmann, O. and Chesbrough, H. (2009). Open R&D and open innovation: Exploring the phenomenon. *R&D Management*, 39(4): 311–316.
32. Torkkeli, M., Kotonen, T. and Ahonen, P. (2007). Regional open innovation system as a platform for SMEs: A survey. *International Journal of Foresight and Innovation Policy*, 3(4): 336–350.
33. Belussi, F., Sammarra, A. and Sedita, S.R. (2010). Learning at the boundaries in an "Open Regional Innovation System": A focus on firms' innovation strategies in the Emilia Romagna life science industry. *Research Policy*, 39(6) July: 710–721.
34. Tödtling, F., Prud'homme van Reine, P. and Dörhöfer, S. (2011). Open innovation and regional culture — Findings from different industrial and regional settings. *European Planning Studies*, 19(11): 1885–1907.
35. Amison, P. and Bailey, D. (2014). Phoenix industries and open innovation? The Midlands advanced automotive manufacturing and engineering industry. *Cambridge Journal of Regions, Economy and Society*, June.
36. Baron, M. (2014). Open Innovation cooperation strategies in Regional Innovation System. *Proceedings of The XXV ISPIM Conference — Innovation for Sustainable Economy & Society*, Dublin, Ireland on 8–11 June 2014.
37. Laranja, M. (2004). Innovation systems as regional policy frameworks: The case of Lisbon and Tagus Valley. *Science and Public Policy*, 31(4) August: 313–327.

## Website Reference for Case Descriptions

http://www.nichevehiclenetwork.co.uk/.
http://www.silesia-automotive.pl/.
https://www.alpine-open-innovation.eu/.
http://www.mobileheights.org/.

http://www.demola.net/.
http://www.chemelot.nl/.
http://www.mobileheightscenter.com/.
http://evertiq.com/.

## Legal Acts and Regulatory Documents

Communication from the Commission, *Europe 2020. A strategy for smart, sustainable and inclusive growth*, Brussels, 3.3.2010, COM(2010) 2020 final.

Position of the Commission Services on the development of Partnership Agreement and programmes in Austria for the period 2014–2020, Ref. Ares(2012) 1283900 — 30/10/2012.

Position of the Commission Services on the development of Partnership Agreement and programmes in Belgium for the period 2014–2020, Ref. Ares(2012) 1335413 — 13/11/2012.

Position of the Commission Services on the development of Partnership Agreement and programmes in Bulgaria for the period 2014–2020, Ref. Ares(2012) 1273775 — 26/10/2012.

Position of the Commission Services on the development of Partnership Agreement and programmes in the Republic of Croatia for the period 2014–2020.

Position of the Commission Services on the development of Partnership Agreement and programmes in Cyprus for the period 2014–2020, Ref. Ares(2012) 1307722 — 06/11/2012.

Position of the Commission Services on the development of Partnership Agreement and programmes in the Czech Republic for the period 2014–2020, Ref. Ares(2012) 1283741 — 30/10/2012.

Position of the Commission Services on the development of Partnership Agreement and programmes in Denmark for the period 2014–2020, Ref. Ares(2012) 1307708 — 06/11/2012.

Position of the Commission Services on the development of Partnership Agreement and programmes in Estonia for the period 2014–2020, Ref. Ares(2012) 1283813 — 30/10/2012.

Position of the Commission Services on the development of Partnership Agreement and Programmes in Finland for the period 2014–2020, Ref. Ares(2012) 1245562 — 22/10/2012.

Position of the Commission Services on the development of Partnership Agreement and programmes in Germany for the period 2014–2020, Ref. Ares(2012) 1320393 — 09/11/2012.

Position of the European Commission Services on the development of Partnership Agreement and programmes in Greece for the period 2014–2020, Ref. Ares(2012) 1337850 — 13/11/2012.

Position of the Commission Services on the development of Partnership Agreement and programmes in Hungary for the period 2014–2020, Ref. Ares(2012) 1286381 — 30/10/2012.

Position of the Commission Services on the development of Partnership Agreement and programmes in Ireland for the period 2014–2020, Ref. Ares(2012) 1320527 — 09/11/2012.

Position of the Commission Services on the development of Partnership Agreement and programmes in Italy for the period 2014–2020, Ref. Ares(2012) 1326063 — 09/11/2012.

Position of the Commission Services on the development of Partnership Agreement and programmes in Latvia for the period 2014–2020, Ref. Ares(2012) 1260385 — 24/10/2012.

Position of the Commission Services on the development of Partnership Agreement and programmes in Lithuania for the period 2014–2020, Ref. Ares(2012) 1273776 — 26/10/2012.

Position *des services de la Commission sur le développement d'un Accord de Partenariat et de programmes au Luxembourg pour la période* 2014–2020, Ref. Ares(2012) 1369418 — 20/11/2012.

Position of the Commission Services on the development of Partnership Agreement and programmes in Malta for the period 2014–2020, Ref. Ares(2012) 1335362 — 13/11/2012.

Position of the Commission Services on the development of the Partnership Agreement and programmes in the Netherlands for the period 2014–2020, Ref. Ares(2012) 1287855 — 31/10/2012.

Position of the Commission Services on the development of Partnership Agreement and programmes in Poland for the period 2014–2020, Ref. Ares(2012) 1138133 — 28/09/2012.

Position of the Commission Services on the development of Partnership Agreement and programmes in Portugal for the period 2014–2020, Ref. Ares(2012) 1309489 — 07/11/2012.

Position of the Commission Services on the development of Partnership Agreement and programmes in Romania for the period 2014–2020, Ref. Ares(2012) 1240252 — 19/10/2012.

Position of the Commission Services on the development of the Partnership Agreement and programmes in Slovakia for the period 2014–2020, Ref. Ares(2012) 1283774 — 30/10/2012.

Position of the Commission Services on the development of Partnership Agreement and programmes in Slovenia for the period 2014–2020, Ref. Ares(2012) 1283858 — 30/10/2012.

Position of the Commission Services on the development of Partnership Agreement and programmes in Spain for the period 2014–2020, Ref. Ares(2012) 1228388 — 17/10/2012.

Position of the Commission Services on the development of Partnership Agreement and programmes in SWEDEN for the period 2014–2020, Ref. Ares(2012) 1320495 — 09/11/2012.

Position of the Commission services on the development of the Partnership Agreement and programmes in the United Kingdom for the period 2014–2020, Ref. Ares(2012)1315758 — 08/11/2012.

Regulation (EU) No. 1303/2013 of the European Parliament and of the Council of 17 December 2013 laying down common provisions on the European Regional Development Fund, the European Social Fund, the Cohesion Fund, the European Agricultural Fund for Rural Development and the European Maritime and Fisheries Fund and laying down general provisions on the European Regional Development Fund, the European Social Fund, the Cohesion Fund and the European Maritime and Fisheries Fund and repealing Council Regulation (EC) No. 1083/2006.

Regulation (EU) No. 1301/2013 of the European Parliament and of the Council of 17 December 2013 on the European Regional Development Fund and on specific provisions concerning the Investment for growth and jobs goal and repealing Regulation (EC) No. 1080/2006.

Regulation (EU) No. 1304/2013 of the European Parliament and of the Council of 17 December 2013 on the European Social Fund and repealing Council Regulation (EC) No. 1081/2006.

# Chapter 11

# LIVING LABS AND OPEN INNOVATION IN EUROPEAN CONTEXT

Bror Salmelin

*Innovation Systems*
*European Commission, DG Communications Networks Contents and Technology*
*bror.salmelin@ec.europa.eu*

The chapter elaborates the background thinking and path for Open Innovation 2.0 conceptual innovation model. It is based on virtual enterprises, Holonic enterprises and fractal enterprises theory, combined with MIT Living Lab concept developed by Bill Mitchell *et al.* Combining this with the internet/connectivity revolution the need to have faster pace and more successful innovation rate led to the thinking of the quadruple helix, including the citizens as active agents in the innovation process, not only as verificators as they were used to be in the previous triple helix thinking. Based on the work of New Club of Paris (Lin and Edvinsson) the structural intellectual capital (IC) is a key for national prosperity. Open innovation integrating the crowd into the innovation process seamlessly seems to increase the structural IC. Hence, integrating all these components: quadruple helix, nonlinear innovation, fractal and dynamic organisations into innovation processes in real world with real market creation with the users who have become co-creators seem to be the key for future success. The new open innovation 2.0 paradigm seems to be serving the innovation needs very well in time — if we dare to take it on board.

*Keywords*: Innovation Ecosystems; Open Innovation; Living Labs; Innovation Models.

## 1. Introduction

European Union has set innovation as high priority as part of the Europe 2020 Strategy. Europe is focusing on jobs and growth through innovation. Innovation Union is one of the key flagships to target this ambitious goal for Europe to become a leading region in the world of modern innovation.

By focusing on both quantitative (3% of GDP) and qualitative goals in innovation policy this has led to a good mix of instruments supporting modern innovation systems.

In the Horizon 2020, framework research and innovation are seamlessly integrated, and entirely new instruments for funding are created. In the text, I will describe those in the context of European Innovation Ecosystem thinking, linking that to the experiences we already have from Living Labs and Open Innovation, since 10 years.

This chapter describes the background thinking and the developed Open Innovation 2.0 perspective on modern innovation Systems.

## 2. Living Labs in European Context

The origin of Living Labs thinking was in MIT where the approach was to construct test and verification environments in laboratory settings to develop and experiment different technology solutions with real users invited to visit those environments. This led to early prototyping with "real users" again with the probability to have faster scale-up of the results.

Bill Mitchell was one of the key drivers in this new research and prototyping approaches.

When discussing that approach from European perspective it became soon evident that from innovation system perspective end-user involvement could be THE key factor for renewing European Innovation System. We have the most demanding but also very diverse user communities for our products and services; how to harness that to increase success rate and speed of the innovation processes in Europe. The Work of Veli-Pekka Niitamo cannot be enough appreciated when developing the strategy but also practising it in large- and small-scale, also in practise.

At the same time, "Democratisation of Innovation" driven by von Hippel triggered the thinking of co-creation and user involvement in the innovation processes.

The industrially led think-tank for Living Labs strategy in Europe was established in liaison with European Commission, DG Information Society in 2003 to conceptualise the European approach. Further, this Living Labs think-tank focused on Open Innovation becoming the OISPG, Open Innovation Strategy and Policy Group.

Soon it became evident that the European approach should be focusing on creation of innovation hubs which would build on the quadruple helix innovation model, i.e., strong and seamless interaction of the industry, public sector, research institutions and universities and finally also the "people".

Target was to create attractive environments which would be attractive for industrial and research investment due to better innovation dynamics. This dynamics would be supported by the public sector and one of the focus areas would be public sector services which could be co-developed with the user communities, in real world settings. Part of this thinking was based on the idea to stretch the boundaries of societal behaviour as well, as we saw the connectivity and ICT-shared environments (with emerging social media) to change the society as well. The quest was to push the boundaries with real world projects including strong technological development too. Only by doing the research and development with citizens we

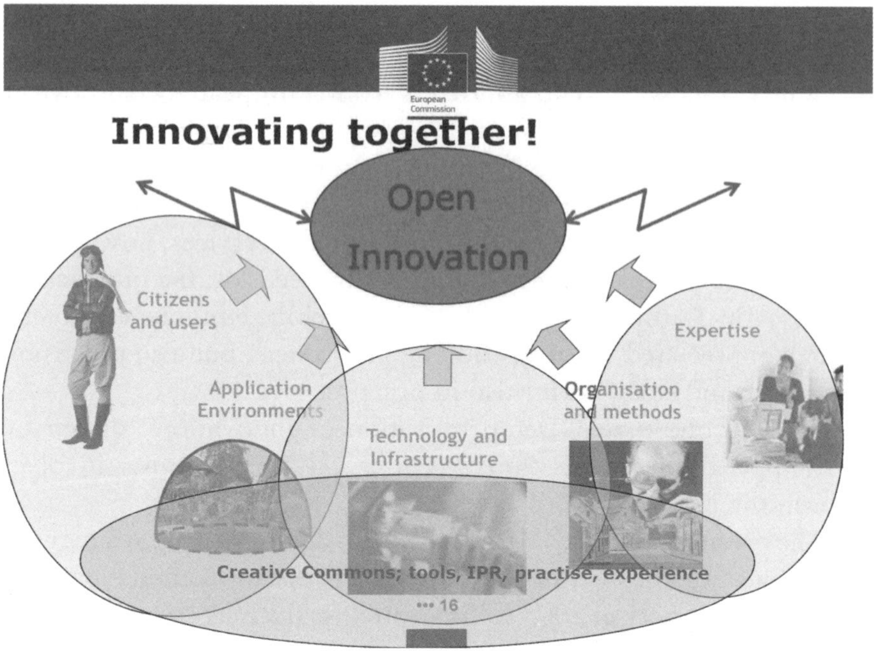


**Figure 1.** Living Labs as European concept

could see what finally would be acceptable and thus scalable to products and services.

This led to the first concept of Living Lab in European context; a real world site, not an extension of laboratory. Importance was set to the scale as it was seen that for scalability we needed the "sample users" to be large enough, at least in hundreds.

In Figure 1, we have all the components needed for European Living Labs: Citizens, application environments, technology infrastructure, organisations and experts. Important to see is the later addition of societal capital into the picture as functioning Living Labs build strongly on the idea of spill-over effects back to the society, giving motivation for all of the stakeholders, including citizens to contribute to the common goal, making Living Lab a winning game.

Based on these conceptual thoughts European Commission and the Finnish EU presidency launched in 2006 the first wave of European Living Labs which built a network, European Network of Living Labs, which became later the ENoLL movement. From the first wave, the network grew fast under the following EU presidencies to the substantial scale it has now, 340 sites even beyond European borders. And, the network is still growing. What we can say that the Living Labs have now a strong foothold in all European Regions, and is being applied as important component in regional innovation systems too.

On European level the networking of Living Labs is of utmost importance. Using Living Labs methodology to find common, scalable solutions with different user environments is essential when driving to common European services based on common architectural approaches. I am happy to see that the thematic cross-border networking of the sites is speeding up, enabling the most interesting Living Labs to collaborate as partners e.g., in the Horizon 2020 projects, especially in smart city or public services context.

## 3. Open Innovation as Part of Living Labs Thinking

As starting point when developing Living Labs in the European way was openness; in sharing platforms for services but also open mindset for collaboration amongst all stakeholders. The thinking stems from the early 90's when the hot topic was virtual and Holonic enterprises which were as group creating both agile and scalable structures for operations; by sharing common operating architectures and by collaborating strongly on task-driven basis. Good examples of Holonic/fractal/virtual enterprise theory was developed e.g., in the IMS (Intelligent Manufacturing Systems) initiative among the leading industrial economies in the 1990's. Scaling up this thinking we come very close to the foundations of Living Labs by adding the public and societal components to it.

Combining the approach by von Hippel about the user-driven and co-creativity in innovation processes with the approach Chesbrough introduced in 2003, open innovation we come to the two fundamental of modern innovation theory. The definition of open innovation by von Hippel focuses on the creation of public goods whilst the one by Chesbrough builds on sharing, cross licensing and in that way being a market- and product-driven approach.

Open platforms, sharing and seamless interaction of all stakeholders is essential in Living Labs. Quadruple helix has thus been central as innovation model from the very first beginning onwards.

Open Innovation Ecosystems are increasingly becoming the synthesis of Living Labs and open innovation processes. We see real new paradigm evolving when combining these. Open innovation has become much more than cross-fertilisation of ideas between organisations, it has become a flow of colliding ideas, raising sparks for new innovations in real world settings.

## 4. Open Innovation 2.0 and Ecosystems

Following the research of Lin and Edvinsson, there are clear indications that intellectual capital, and especially structural intellectual capital drives competitiveness and innovation. This means in turn that from innovation policy perspective the interaction fluidity is a critical feature of any successful innovation system.

Fluidity in this context means frictionless interaction, experimentation in real world, and a lot of unexpected, non-planned collisions of ideas, problems and of course competencies to collide, giving the spark. It is not only about single excellent components in the system, it is centrally about collisions and connectivity.

It was already shown in 2004 that the diversity of research teams increases significantly the probability of breakthroughs, and actually we can also say that mediocre inventions are not enough. We need to combine the best. Cross-fertilisation of ideas is nothing new as such, but what ecosystem thinking does is embedding diversity and serendipity in the innovation process more systematically than ever before.

It is important to move from clusters to ecosystems in our innovation system design. It's nothing wrong with clusters, but they tend to be rather monolithic focusing on one sector only. Of course the clusters reinforce the sector they work in, but the tendency towards improving, extrapolating than to create something new. Hence the emphasis on modern innovation systems need to be increasingly on the "in-between" areas where creation of new is likely, and as consequence also the fast growth.

To substantiate the potential for new market creation the end-users need to turn to be active drivers together with the other stakeholders in jointly creating the new. Quadruple helix innovation model gives clear roles to all stakeholders, including the users as active agents from the first beginning. Earlier the users were objects in the process, not co-creators. By taking the users actively on board we see immediately which solutions can be scaled up and which will fail due to various reasons. Scaling up fast the emerging successes is a key to maintain the dynamics in the innovation system. There are also indications that those organisations which cut failing projects at earlier stage will be more successful in the longer run.

Again, we come to the ecosystem when we think about where the experimentation and early prototyping is to be done. In real world settings, one can at early stage see the potential and also identify the paths for fast enter into the full-scale markets. Seamless user involvement is thus essential. It is important also to understand that properly designed innovation ecosystems provide a safety net from the ideation to the market. Failing fast means also often failing small, and experimentation and early prototyping in turn means faster results to be brought to the market, even incrementally.

Business model experimentation in these open innovation ecosystems is also essential. Due to the dynamics in the economy and technology, it is rarely possible to write the old-fashioned extensive business plans. Often it is enough to have a business model idea and develop it continuously further in the real world settings, to finally see what works and what not. Fast adjustment and experimentation is the way forward.

Here legislation can play also a remarkable role if it is a catalysing one. Restrictive legislation again is a strong hinder for business model innovation. Proper legal framework is one of the important factors for the fluidity of the innovation space we spoke about it earlier.

Innovation has moved from linear processes to mash-up processes where diversity, speed and experimentation are the fundamentals.

We have moved from closed innovation to open innovation and further towards open innovation 2.0 which highlights the interaction, fluidity and mash-up nature of innovation processes, including all stakeholders in quadruple helix innovation.

Open Innovation ecosystems can be regional or thematic, or both. They are built on strong interaction between the competencies illustrated in Figure 2 by different coloured dots. The ecosystem itself has tens or more projects (funnels in Chesbrough sense) which can be more or less leaking to broaden the competence base of each action. The funnels in this context represent development projects, not organisational boundaries. Spill-over effects to the whole ecosystem

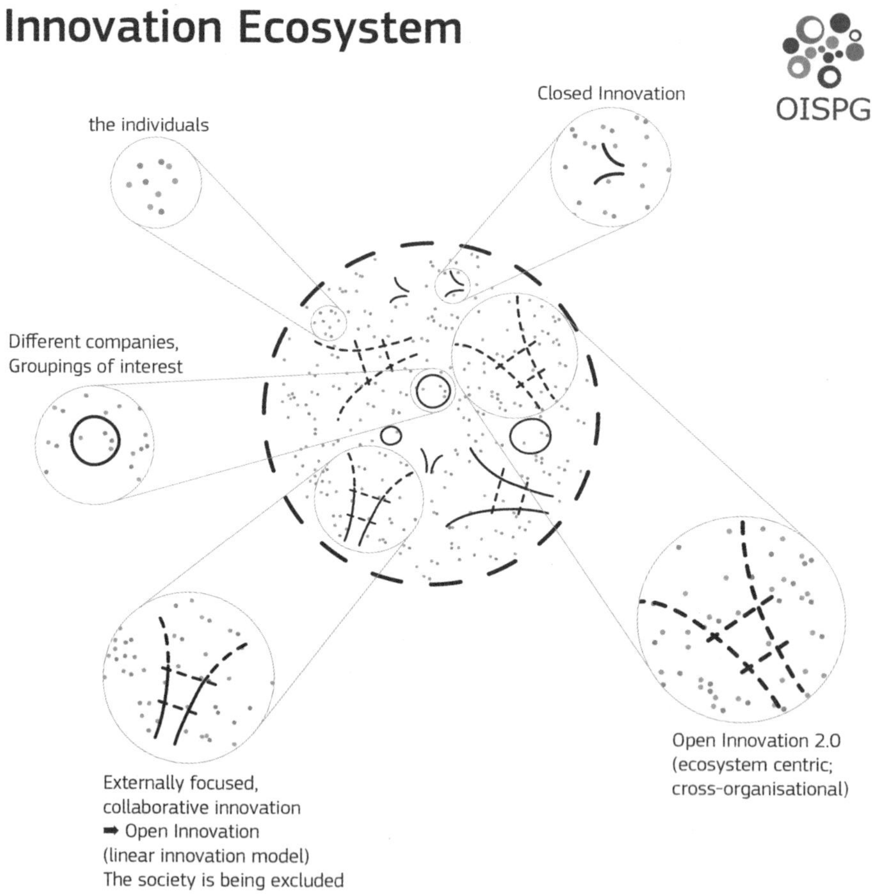


**Figure 2.** Open Innovation 2.0 view on open innovation ecosystems (OIE)

by the projects should increase the societal but also knowledge structural capital enabling continuous rise of the value proposition of the new activities.

Sharing infrastructures but also experience and knowledge is a key of trust to be built within the ecosystem itself. The trust is of very high importance because of the interdependence of all stakeholders in this mutual win–win process. Cross-fertilisation and sharing does not happen without trust.

In ecosystems it is, as previously stated, very important to allow collisions to spark the real innovations, even disruptive ones. Hence the creation process requires courage to design a governance structure for the ecosystem to let it grow organically. Prototyping and experimentation of policies is one of the important components in this development too.

## 5. Open Innovation 2.0 in Twenty Snapshots

In the white paper from 2013 manifesting the Open Innovation 2.0 paradigm Martin Curley and Bror Salmelin highlight 20 key elements as the transformative factors for the modern innovation approach.

The OI2 approach emphasises the importance of Quadruple Helix innovation where the private, public and research institutions collaborate seamlessly and from the very beginning the user(s) communities co-create the new products and services. This leads to win–win approaches as the users get products and services they need, and the suppliers get scalable products and services. If this co-creativity and prototyping in real world settings would not take place there would be a real risk that the development work would lead to a win–lose setting between the existing players in the market, and no new markets would be created either.

Cross-disciplinary innovation together with prototyping and experimentation is bringing forward the required dynamics. Failing fast and getting directions to potentially successful solutions at early stage is essential. Traditional piloting or test bed approaches are not sufficiently scalable to verify the market potential of the inventions.

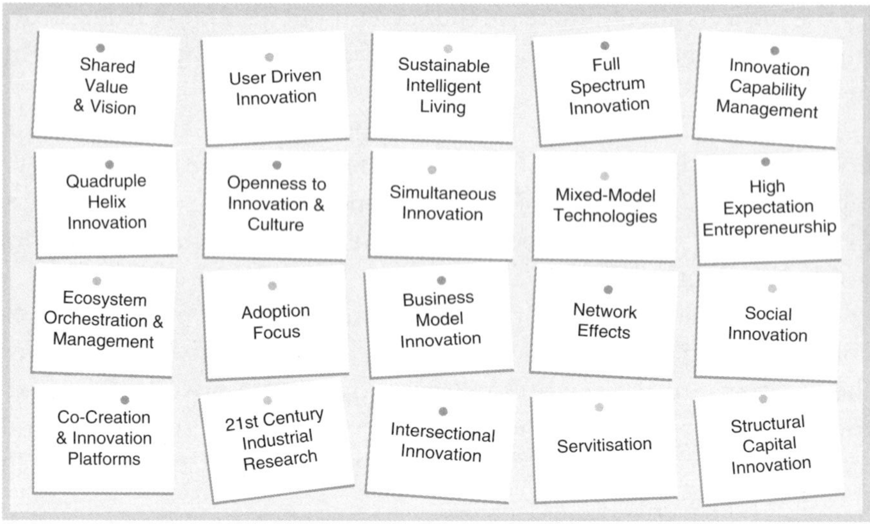


**Figure 3.** Twenty drivers for Open Innovation 2.0

In this palette of 20 drivers for Open Innovation 2.0, (Figure 3) one needs to highlight both societal and technological innovation which enable business model (more generally value creation model) innovations. The area of business model innovation together with the new markets emergence are clearly dimensions/realms in which we Europeans can do/perform much better.

How to achieve the fluidity and frictionless environments for multi-stakeholder trials, including legal and policy elements is the key to root in the European mindset. We need to speak about openness for innovation, innovation 2.0 culture, to complement the view.

## 6. New Types of Leadership, New Processes and New Approach to Ecosystems — Paradigm Change is Real

The paradigm has changed. Figure 4 illustrates some aspects to concretise this change and illustrate its drivers. Of course, these factors are interrelated in complex systemic manner and lead to the need of

| *Closed innovation* | *Open innovation* | *Open innovation 2.0* |
|---|---|---|
| *Dependency* | *Indepencency* | *Interdependency* |
| *Subcontracting* | *Cross-licensing* | *Cross-fertilisation* |
| *Solo* | *Cluster* | *Ecosystem* |
| *Linear* | *Linear, Leaking* | *Mash-up* |
| *Linear subcontracts* | *Triple Helix* | *Quadruple Helix* |
| *Planning* | *Validation, pilots* | *Experimentation* |
| *Control* | *Management* | *Orchestration* |
| *Win–lose game* | *Win–win game* | *Win more–Win more* |
| *Box thinking* | *Out of the Box* | *No Boxes!* |
| *Single entity* | *Single Discipline* | *Interdisciplinary* |
| *Value chain* | *Value network* | *Value constellation* |

**Figure 4.** The change and drivers of the innovation paradigm

looking at successful innovation ecoSYSTEMS and innovation processes together.

Closed innovation reflects the traditional linear paradigm, often based on brilliant individuals or performing industrial labs. Open Innovation, as introduced by Henry Chesbrough, is a move towards collaborative innovation structures, where those ideas not used by oneself can be seen as tradable assets to those who might have need for specific technologies.

When we began to analyse the innovation processes and the success closer, we realised that one of the critical elements is the scalability of the work, which naturally results in increased success rate. But how to achieve this?

We need to break out from the traditional linear models; we need to dare to do more experimentation in real world settings as only then we learn very fast what is scalable, successful, as opposed to what is simply not worth going forward with. Traditionally, we see pilots and validation in many projects, but ... often they come too late to have any influence of the project work itself. This triple helix approach which excludes end-users from the actual innovation process is by far too slow. Only by moving to the quadruple helix model where the innovation process happens “out there” with real people in real environments we can speed up the successful results and kill the bad ones on time.

Another dominant element of the open innovation traditional cross-licensing process is the cluster thinking. Cluster operations reinforce well the competitiveness of sectors. However, the challenge is not only to stay competitive in the existing field, but also to find entirely new areas for value creation. We need to have interdisciplinary manner actions between the clusters in the open innovation ecosystems to strengthen cross-fertilisation. And, taking the users on board and integrating them into the innovation process from the very beginning will lead to the creation of new markets. If we target only traditional clusters and traditional industries we easily end up with a win–lose game.

Organisational changes and collaboration changes are also clearly moving towards this mash-up, mixed disciplines approach. Value chains with subcontractors highlight the linearity in innovation processes together with control approach which is typical for the manufacturing and traditional industry era. When products integrate into services and get more complex, we have seen networking between suppliers to be established, e.g., in the automotive sector, where independent component manufacturers deliver to many brands simultaneously, based on their special competencies. In open innovation 2.0, we go even further into the dynamic value constellations where the links are not *a priori* determined, but more task driven. Competencies and resources are combined based on the tasks, not as earlier when the services were determined by organisational structures. In turn, this also means that the end users will be much more dominant in the innovation process for modern products and services, especially on their functional level.

The innovation process change affects also radically the management styles of successful companies. We have plenty of examples where an authoritarian control-type of management is replaced by strong leadership. However, we need to go into even further metaphors when we move to open innovation 2.0. The successful leadership will be mentoring, catalysing, inspiring: It will be orchestration of fluid resources to perform their best. And, what makes all interesting is that the orchestration conducts not only the known players, but also the audience to create fantastic joint experiences with the interaction internally AND externally; Like in a successful concert where the ambience and success is all about the interaction and not just the play, even professionally.

Open Innovation 2.0 is a new mindset; it is openness for innovation. It is the courage to experiment and prototype. It is the courage to fail and scale. And, as a consequence, it builds up a growing spiral of performance built on success and motivation.

## 7. Conclusion

Living Labs, networked society, democratising innovation, open innovation, disruptive innovation ... Many words which are fluently used without often thinking about the reality behind. The reality is however in the courage to change the behaviour, including the governance structures to create something new. The reality is also to turn these buzzwords into a functioning innovation ecosystem with new dynamics.

ICT provides connectivity and the shared space of knowledge, meaning that the new paradigm of open innovation ecosystem is more doable than ever before. In the rich connectivity, we need to see the new role of all players in the spirit of quadruple helix innovation, and move due to the dynamics needed to an experimentation and prototyping culture. This shows the options for success earlier and significantly reduces the risk for big failures too.

The fundamentals are developing in Europe. Our challenge is to make these fundamentals to work together, to fully use the potential we have as the single biggest market in the world.

The paradigm has changed.

# INDEX